William W. Walker

An Itinerant in the British Isles

William W. Walker

An Itinerant in the British Isles

ISBN/EAN: 9783743406889

Manufactured in Europe, USA, Canada, Australia, Japa

Cover: Foto ©berggeist007 / pixelio.de

Manufactured and distributed by brebook publishing software (www.brebook.com)

William W. Walker

An Itinerant in the British Isles

AN ITINERANT IN THE

BRITISH ISLES.

BY

REV. W. W. WALKER.

TORONTO:

WILLIAM BRIGGS.

C. W. COATES, Montreal. S. F. HUESTIS, Halifax.
1896.
71

INTRODUCTION.

WHEN there are so many books of travel clamoring for the public eye, there may seem a degree of temerity in venturing another into the arena. We do not claim for our modest little volume any distinctive merit. We do not hold it up as a model. The reader may find much to criticize, but we hope he may find something to interest and please, and we crave his indulgence. Our only excuse for the publication of the little volume is the desire to preserve in this form the record of a visit to the " home land " so pregnant with pleasure—a desire stimulated by the requests of friends for copies of the book as soon as in print.

We have sought to describe the scenes visited, and to record our impressions so as to give those readers who have not had the good fortune to cross the Atlantic, a faithful picture of the experiences of travel. We have spared no pains to secure accuracy in every detail, and if errors have crept in, it has been due to misconceptions or because we were wrongly informed in the matter at issue. To the Canadian a visit to the Old World opens up a mine of inexhaustible interest, and we wish our readers, one and all, the pleasure of a tour through those "sea-girt isles" that bear to us the tender relation of the Mother Country.

THE AUTHOR.

CONTENTS.

AN ITINERANT IN THE BRITISH ISLES.

CHAPTER I.

MONTREAL AND MAINE.

WE left the Queen City of the West on the twenty-ninth day of April, eighteen hundred and ninety-four. After travelling all day through familiar scenes, we reached that great commercial metropolis, Montreal, about an hour after dark; and having some time on our hands before securing berths on the train which was to carry us farther east, we were not long in discovering the fact that the city had wonderfully improved since our college days. A magnificent and imposing structure now marked the site of the old Bonaventure station, the new edifice being a credit and an ornament to this great and substantial city with a quarter of a million inhabitants.

We had taken a palace car from Toronto, and now engaged a berth in a sleeper for Portland, United States having had to remain until late at night at this important head of ocean navigation, but at last our train slowly moved out from the depot, having on board a number of persons bound for the Atlantic seaboard, amid partings such as chill the life from out young hearts.

A Montreal gentleman had placed his wife and two children on board, their destination being Scotland. The parting between husband and wife was too sacred to mention in detail, but as the motion of the train slowly increased, the lady took one last look out of the window in the direction of the husband whose form was gradually receding from view, and he, noticing the movement, waved a last farewell, which was too much for the faithful wife, who turned to us with tears in her eyes, exclaiming, " Oh, I wish the train would stop. I cannot leave him. I cannot leave him alone !" But the train would not stop, and she soon settled down, and eventually became a brave and resolute woman, both by land and sea.

We cannot, of course, describe the country which we passed through during the hours of darkness, although but few slept—what with the rocking of the train, whistling of the locomotive and universal din and noise. Even though our wearied bodies invited it, yet " nature's sweet restorer, balmy sleep," was practically impossible, as very many discovered with sorrow, the writer included. In spite of a sleepless night, however, we were astir at daybreak, and discovered the fact that the grand old Union Jack no longer floated over us, but wherever a flag was to be seen, it was the Stars and Stripes, reminding us that we were now on the territory of Uncle Sam. We were not prepossessed with the appearance of things generally, as we traversed the State of Maine toward the ocean. The buildings were mostly wooden structures, and seemed unsubstantial and cheap-looking, and not at all in keeping with the assumption to greatness made by the American people; and yet in one way they have put us Canadians to shame, for in this

State they have enacted a prohibitory liquor
law, and we must in this give our American
cousins the palm, and pray that our golden age
of Prohibition may soon be ushered in.

As we journeyed on, the country became
mountainous — hill and dale, mountain and
valley alternating. Some of the hills or mountains were of considerable height, and their
sides were clothed with forest; and as the season
advances and the trees burst forth into foliage
we doubt not that the scenery will be gorgeous
in the extreme, and could not help feeling that
a hardy race and brainy and virtuous people
were cradled among those everlasting hills, and
their temperate habits and prohibitory liquor
law proves the assumption to be correct; but in
spite of all this, as we viewed the public edifices
in the towns and villages through which we
passed in this eastern State, we could not but
feel their inferiority to our Canadian towns and
villages with their public buildings.

Sometimes in order to appreciate aright our
national heritage, it is necessary for us to go
abroad, and in other lands we learn to love more

and more the land which God has given to our fathers as a heritage.

It was now toward noon, and we were thundering along with great rapidity over threads of throbbing steel towards the mighty Atlantic. At last an exclamation burst from the lips of someone in our car that made us all start, " The sea, the sea !" and for the first time in our lives some of us caught sight of the briny waters of the mighty deep.

A few minutes more and we were in the depot. Our first impressions of Portland were, however, not very favourable. As we walked about and viewed more of the city and public edifices, we began to realize that in this case first impressions were going to prove lasting.

We sought out the Post-office and procured several cards to send to friends in Canada, letting them know of our safe arrival at the sea-coast. At the Post-office they refused to accept Canadian money, and when we tendered them their own money in payment for the cards they gave us change in American cents—worthless, of course, any place else—which we have to this

day, reminders of our short sojourn in the sea-
port of Maine.

We had spent some time a few years before in
one of the western States, attending a school of
elocution and oratory, and found the American
people extremely courteous and friendly. We
thus found it hard to account for the discourtesy
and selfish independence of the Portland people,
but fortunately we were not compelled to endure
it long. As it was announced that our steamer
sailed at one o'clock, we found it necessary to
repair without further delay to the docks, and
oh, such docks! We had expected to see neat
masonry, but instead found wooden wharves
which seemed so high above the water—we
thought thirty or forty feet—the tide being out.
They would not have looked so ridiculous if
the posts or piles had been perfectly plumb,
but they were twisted about in all shapes, and
as we inspected them, we may say that we were
simply disgusted.

We now found that for some unexplained
reason the steamer would not sail until four
o'clock in the afternoon, giving us a better oppor-

tunity to acquire information. The day was all that could be desired for this purpose, a glorious May Day, the first of the month.

We were anxious to see an American crowd, that we might make observations with regard to their physique and general appearance in comparison with Canadians, and this piece of information was also furnished gratuitously, for on the arrival of the hour for sailing, about one thousand people had assembled on the docks to witness the departure of our good ship the Royal Mail Steamer *Labrador*, of the Dominion Line.

In stature, the Americans are scarcely as tall as Canadians, and as a general rule we feel safe in saying that an American of average type does not represent as much matter avoirdupois as a Canadian. We particularly noticed this some years ago on the parade ground at Niagara, where many of the American troops from the fort across the river had come to witness the evolutions of our boys. When manœuvring was ended, they mixed indiscriminately for a time, and we noticed that the Canadians were considerably larger than the Americans.

CHAPTER II.

LEAVING PORTLAND—LIFE ON THE ATLANTIC OCEAN.

FOUR o'clock had now arrived, and our observations on shore had come to an end for the present. The powerful steamer slipped from her moorings and soon the ponderous machinery was in motion, and now farewells commenced. The crowds on shore waved their hats and handkerchiefs, which farewells were returned by signals, and the discharge of rockets from the ship.

A Portland steamer accompanied us out for some distance, and then after touching with her prow the prow of our steamer, and dropping her flag in farewell, she turned about and steamed for shore, leaving us alone in our course toward the rising sun.

The first thing of interest we passed, at the mouth of the harbour, was the battery, composed of dynamite guns protecting it, and although no friend to war, we always felt a great interest in studying fortifications and arms. As far as we could judge from the deck of the vessel the way the guns were trained and laid, and the utter absence of adequate earthworks and protection generally, a hostile fleet armed with rifled guns of heavy calibre, and standing well out to sea, would not need to expend many projectiles to dispose of the gunners and dismount the guns, and subject Portland to all the horrors of a bombardment.

As we left the mouth of the harbour and got fairly under way upon the bosom of the Atlantic, the motion of the vessel perceptibly increased, and with this came that all-absorbing sensation known as seasickness. This seems to be, to some extent at least, a nervous trouble, induced by imagination. As proof thereof we heard from good authority of one lady who became seasick before she got on board the ocean liner, from sheer imagination. The increase of sickness proved that although much of it is nervousness and induced by imagination, yet the rising of the vessel to the crest of the

2

wave, and its falling into the trough of the sea, causes everything to have an upward tendency, and ere long many a good dinner was lost to poor suffering humanity.

It moves the coldest nature to sympathy to witness the abject appearance of those who are suffering from this malady. They look as though they did not possess a friend on earth, and did not care whether they lived or died. Things would undoubtedly have been worse if we had been on board a vessel that was subject to heavy rolling, as there is a vast difference between ships in this respect.

The *Labrador* did not labour in the swell as some vessels would.

A description of her size will be in place, perhaps, at this juncture. Her length is four hundred feet and her beam forty-seven. She is four thousand seven hundred tons register, draws twenty-five feet of water, her engines being of seven thousand horse-power, and her speed fourteen knots an hour. She consumed sixty-five tons of coal per day, and had four masts of hollow iron with steel rope rigging, and was in every respect a powerful and substantial craft, with good and comfortable accommodation in both cabins and steerage.

It was said by some of the passengers that the British Government had a claim upon her, so that in the event of war she could be turned into a transport after having been armed with a few quick-firing guns.

It was also said that she carried several men of the Royal Naval Reserve to facilitate matters in the event of the breaking out of hostilities. This, of course, we could not vouch for, but one thing we do know that her captain and the full complement of her officers were capable, courteous and efficient men. We especially thought that the spirit of the captain was that of a true gentleman, and all the sailors, as far as we could see, were a well-behaved, good-hearted lot of men, except, perhaps, occasionally some of them indulged in language that was in the eyes of a theologian a little unscriptural.

One sailor was up in the rigging, and while making some repairs, perhaps twenty or thirty feet above the hurricane deck, used the term

"Hell" with a good deal of emphasis when anything went wrong. Having become well acquainted with him, and finding him a smart, intelligent young fellow, we suggested from the deck as we stood underneath, that he use the term "Hades" instead of "Hell," as it sounded more theological. Instead of becoming angry at the suggestion he laughed most heartily, and while completing the work we did not hear him use any more scriptural language in an improper sense.

We had always heard that sailors were a very wicked class of men, but must say in all candour that we did not find them exceptionally so, but on the contrary found that despite their faults they are a good-hearted, jolly, happy-go-lucky lot of fellows. Their wit also is surprising, as exemplified by the following: On one occasion a sailor while working in the rigging dropped his marlin spike, a thick iron rod about twelve or fourteen inches long, sharpened at one end and with a hole punched through the heavy end, through which is passed a cord which seamen hang around the neck. For a moment our friend was staggered as to the course he should pursue, as doubtless the captain would be displeased because of the loss of the instrument. At last, however, he went to the captain and said, " Captain, is anything lost if you know where it is ? " " Certainly not; what nonsense ! " said the master of the ship, impatiently. " Well, Captain," said the sailor, " my marlin spike fell overboard, but according to your ruling it is not lost, as I know that it has gone into the Atlantic Ocean." The commanding officer instantly turned on his heel and disappeared; his own logic had been applied and he dare not contradict it.

We had not been out more than a couple of days from Portland when we sighted a school or shoal of porpoise sporting in the dark blue waters. With regard to their number, we may say that their name was legion, for they were many. It was extremely interesting and inspiring when six or seven hundred miles out at sea to notice so much life in the waters. The seagulls also followed us, and hovered on swift wing around the steamer continually. The presence of so many winged creatures of extreme beauty in many cases, and also of so much fish life made existence on the watery main anything but lonely.

A little later, when the interest created by porpoises had somewhat subsided, an enormous iceberg hove in sight, soon another and another, until we had seen quite a number. We always had the impression that the outside of the icebergs would be perfectly clear and glassy, but such was not the case. They are white as snow and of indescribable beauty; the whiteness is accounted for by the action of the sun on the huge masses of ice. We were particularly anxious to know the height of some of the towering masses, and were informed by the ship's officers that one at least was two or three hundred feet clear of the water, and when we consider the fact that but one-third is above the surface and two-thirds underneath, it enables us to form some conception of the enormous size of one of those masses, some of which look like floating polar palaces.

One especially, with its towers and pinnacles,

looked like the ruins of some grand old marble castle.

Some, however, are so perfect in their formation and so symmetrical that they do not at all resemble ruins.

One of mighty proportions and cloud-piercing height and transcendent beauty reminded us of what we imagine the great white throne of God in the heavens will be. Nothing on our voyage seemed to inspire us with such sublime thought and with so much awe as the sight of those stupendous masses of glistening white.

We sighted many vessels at sea. Some were fishing off the coasts or banks, more properly, of Newfoundland; some were large, full-rigged sailing ships with three masts. The most stately vessels we were privileged to sight at sea were the full-rigged, three-masted, unarmoured ships. They are of such beautiful symmetry and ride so majestically.

We also met some large ocean liners steaming along rapidly towards New York and Portland, and a little later someone cried out, "A whale! A whale!" and about half a mile away we could see him spouting, and, as the vessel drew nearer, frightened doubtless by the noise of the machinery, he lifted his huge head about fifteen feet out of the water and disappeared. His mouth was of prodigious dimensions, and the jaws set with murderous spear-like teeth of enormous size. One enthusiastic passenger who had been rather sceptical as to the truth of the Bible story of a whale swallowing Jonah, said that he would never doubt the Bible story again, after seeing the open jaws of an Atlantic whale. The sensation .caused by the appearance of the great marine monster was the last of a week of surpassing interest and profit.

Soon the holy Sabbath dawned upon us. It was the custom to have divine service on board every Sunday morning, and as the captain of our steamer belonged to the Roman Catholic Church, the task of assembling the cabin passengers in the saloon for worship was delegated to the surgeon, who also conducted the service of the Church of England, required to be read on all Her Majesty's ships.

The surgeon, who was a man of splendid

physical presence and gentlemanly manners, invited the writer to take charge of the service. We protested that we were not familiar with the service of the Church, but he insisted that he would sit behind us and give direction as to the passages to be read and also the prayers for a morning service at sea; and then to help us to decide he suggested a short sermon at the close of the ship's service.

We could not resist further, and as a precautionary measure, because of a considerable motion of the vessel, we drew a table so close to the side of the saloon that there was just room for our lower limbs, and wedged in that position, with nearly all the saloon passengers and many from the second cabin before us, and our friend the surgeon at our side, we proceeded with the service, and to our infinite relief got through the entire performance without a single hitch. The danger of bungling was now past, as we felt perfectly at home in the delivery of a Methodist sermon. A Catholic priest who had been holding service in another part of the ship for the benefit of the passengers of that persuasion, returned in time to hear the sermon, and having become well acquainted with him, finding him intelligent and cultured and principal of a seminary in Montreal, we felt flattered as he complimented us on our eloquence. The text of the morning had been, "I am not ashamed of the gospel of Christ by land or sea." He assured us again and again that he very much enjoyed the discourse. This encouraged us, as we felt that, what with the lurching of the steamer, the roar and throbbing of machinery and the strangeness of a religious service at sea to one who had seldom been off *terra firma*, the sermon was almost beneath contempt.

In the afternoon, the steward assembled the intermediate and steerage passengers in the second cabin, and asked us to address them, which we cheerfully did, but the roar of machinery was so great here that we had a difficult time in the delivery of our sermon, but hope that some soul was refreshed and blessed in spite of the weakness of human agency.

The remainder of the Sabbath passed quietly away, and with it the beautiful weather which we had so much enjoyed since leaving home. There were now signs of an approaching storm; we had always thought that we would like to see a storm at sea, and the desire of our heart was about to be realized.

The wind, which had hitherto been conspicuous alone for its absence, now continued stiffening every hour, and the motion of the huge mass of iron which was our home for the present, continually increased. This condition of things went on until at last some oak rollers, which were being carried as a part of the cargo to Liverpool, and which had been improperly stowed away in the hold, broke loose and commenced rolling from one side to another, and every few moments, as the vessel lurched, would strike broadside with hideous crash. Some were afraid that perhaps they would go through the side of the ship.

As darkness came on the storm increased, and about ten o'clock the vessel pitched so heavily that we thought we would like to go on deck and view the wild scene. We had, however, scarcely reached the top of the stairway, when the bow went down into the trough of the sea. As the stern was on the summit of the retreating wave, the result was that the vessel shipped a sea, and as the water rushed like a river down the deck toward the open hatchway, we were glad to slip back to our stateroom and keep quiet for the remainder of the night.

Few of the passengers slept during the eventful hours of darkness. Every few minutes during the night there was a fearful crash. We tried in vain to determine as to the cause, and in the morning asked one of the officers what it meant, and he said it was caused by the vessel shipping seas and the water dashing down the open hatchways of the upper deck, and as it struck the second with great force, causing the crash described above. The hatches on the next deck were all closed.

During this day we had a magnificent opportunity of watching the waves, although it was extremely difficult to keep one's feet, as all day the vessel rolled fearfully.

We had often heard of the waves rising to a certain estimated height, and now carefully endeavoured to measure the height by the eye, and by the number of feet between the water-line of the ship and the main deck, and also by asking the officers and sailors.

The result of the estimation was that many of the waves from trough to crest were fully thirty feet.

Having often read in poetry of the waves rising mountains high, on looking all around over miles of ocean lashed into foam by the fury of the storm, we could not but feel that the descriptions heretofore given of the waves were a gross exaggeration. We would say, and certainly with more truthfulness, that in a storm on the Atlantic Ocean the waves rise like little hills, and in writing thus, the sublimity and grandeur of the awful scene is not lessened.

There is perhaps no place on earth where men can get such correct conceptions of the power of the everlasting God as in an iron-clad steamer of nearly five thousand tons, tossed like a cork on the foaming crests of Atlantic billows.

Well might the inspired servant of God say,

" They that go down to the sea in ships, these see the works of the Lord, and his wonders in the mighty deep." Speaking of the wonderful works of God, the Psalmist says, " There go the ships: there is that leviathan, whom thou hast made."

Note he does not here speak of the work of men's hands as material ships, but as he speaks of leviathan which God has made, he uses it in the same sense as the *nauticus* (ship), which, instead of meaning any fabrication of man, means a creature with wings set like sails, so that the wind, catching the sail-like wings, propels it over the bosom of the waters; the workmanship of hands divine, in both cases alike showing, and intended by the Psalmist to show, the power and skill of Him who holds the waters in the hollow of His hand.

After studying God amid the wonders of the deep, and viewing nature in her varied hues, and behind her discerning a supreme something that rules a universe, the object of man's adoration; also, after experiencing to the full the restfulness of ocean travel, we began to long for a glimpse of shore, even though it should be but a passing one. Ship-life, being at first a novelty, is exceedingly pleasant, but gradually the pleasing novelty passes away, and the eyes grow weary of continually looking out over an apparently shoreless waste of waters.

CHAPTER III.

VICTORIA ISLAND—THE COAST OF IRELAND.

THE desire of every heart was gratified by a glimpse of Victoria Island, or as the sailors familiarly call it, Tory Isle, dimly outlined on the far-off horizon, and as we drew nearer we could discern nothing particularly interesting or attractive in the appearance of the place, but to many because it was land, it appeared a veritable paradise, especially to those who had been sick nearly all the way.

Soon, however, this place was forgotten in the sight of something of infinite and indescribable beauty. It was nothing other than the coast of Ireland, its hills rising almost from the water's edge, with but few habitations nestling among them, but of romantic beauty,

We could not but feel that the picture of this historic isle, though artistically fashioned, had not been overdrawn, and as our gallant ship

sped on her way, leaving behind many miles of
coast with hill and dale and lovely cottage,
many regrets were expressed that she was so
rapid a sailer. It would certainly have been
delightful to have lingered on such scenes with
all their enchantment. The sublime words of
the sacred bard were suddenly recalled, "my
willing soul would stay;" but with man's natural
hopeful expectancy we looked forward to visions
of still brighter scenes, and truly more glorious
sights were yet to follow.

The world-famed Giant's Causeway was soon
discernible to the naked eye. We passed quite
close to this towering wall of rock. It was, as
nearly as we could judge, between two and three
hundred feet in perpendicular height, with its
face fluted like the cylinder of a well-constructed
revolver. The rock was greyish in colour, and as
far as we could observe from the deck, was not
the Laurentian or oldest formation, but probably
of the Cambrian period of geology.

After passing this object of interest the country became more level, and beautiful fields and
farms were to be seen on every hand. Both the

3

farms and divisions and sub-divisions thereof
seemed to be perfect squares; the fences, which
were composed of stones in some cases, and in
others hedges or ditches, were as straight as a
gun-barrel, and the farm-houses or cottages
seemed to be all either painted white or white-
washed, everything evidencing thrift and clean-
liness, and the fields, oh, what a glorious green !
We were again reminded of sacred hymnology,
where the poet, in ecstasy, speaking of the
heavenly country says,

> " Sweet fields arrayed in living green,
> And rivers of delight."

We did not see many rivers, but truly the
green of the fields was a living green, and the
waters around the island seemed of a greenish
hue, more beautiful than anything we had ever
seen in that colour before.

As we noticed all this, we could not but think
that the term " Emerald Isle " or " Green Isle "
applied to Ireland is most fitting and appro-
priate. Everything is green in connection with
it except its people, and from our experiences

with them toward the close of our travels, we
shall endeavour to show farther on in this work,
that they are thoroughly seasoned.

We noticed some beautiful homes in the north
or north-east as we continued on our course ;
fine houses, built in an attractive style of
architecture, and situated in lovely parks, shaded
by the overspreading branches of lordly trees,
many of them doubtless centuries old.

The landlords, or land-holders, of Mono's Isle
have homes that will, in stately magnificence,
equal, perhaps, any upon earth.

As we gazed upon the scene that passed in
panoramic view before us, far away towards the
interior of Ireland we descried ancient round
towers situated upon lofty hills, substantially
built, reminders of a past full of significant
interest ; also, alas ! alas ! full of tragedy untold.
But we must not dwell on that which is hidden
behind the monuments of the far past, with its
too often grim memories of scenes that perhaps
are better consigned to the shades of oblivion.

After having passed the bay that opened up
water communication with Belfast, the most pros-

perous and wealthiest city in Ireland, we spent
some time in meditating upon what we had seen,
and the result of this meditation was, we pro-
nounced this wonderful island the brightest gem
in Victoria's crown.

CHAPTER IV.

SCOTLAND—THE IRISH SEA—THE MERSEY RIVER.

EXPATIATING on the glories of the emerald of the sea, we have overlooked the land of brown heath and shaggy wood, land of the forest and the flood; in a word, we have been silent about bonnie Scotland, and after passing miles and miles of its coast we must endeavour to describe its general appearance.

The entire coast that we passed seemed to be mountainous, and it surprised us to find the channel so narrow that separated Scotia from Old Mono. On seeing this, however, we were easily led to understand how the Scottish people and the inhabitants of northern Ireland had a very similar dialect, their accentuation, that is, of those on the adjoining coasts, being precisely the same. This is readily explained, when we consider the fact that in trade and commerce they were continually intermingling.

The western shores of Scotland, as discernible from the sea, resemble, to some extent at least, the western part of Ireland that we first sighted after passing Victoria Island.

It was here, cradled among those everlasting Scottish hills, that one of the most vigorous races, physically and mentally, that the world has ever seen was nurtured.

It was from among those very slopes upon which our eyes now gazed that men, hardy, intelligent and brave, went forth to fight their country's battles, and almost invariably to win. It was from among those mountain scenes that men, with philosophical mind and analytical brain, went forth to occupy chairs in institutions of learning and charity. It was from those rugged scenes that men, whose boyish feet had once pressed the soil of hill and dale in mirthful glee, went forth to sit in senatorial chambers and legislative halls, to enact laws and frame codes and constitutions that were destined to regulate the actions of millions of their fellow-creatures.

No marvel, then, that in battle charge her

sons have been fired with a noble patriotism, and have swept forward to honour and triumph with the words "Scotland forever!" trembling upon their compressed and bloodless lips. No marvel that in the frozen regions of the far north, and on the plains of the sunny south, are men who render unto her to whom it is due a splendid tribute of devoted love.

Night once again threw her sable mantle over land and sea, and, as we retired to our state-room for slumber, a strong breeze sprung up which soon freshened into a gale, and our experiences on the Irish Sea became anything but pleasant, except, perhaps, for the knowledge that we would so soon be through our ocean voyage.

All night long we were tossed in the cradle of the deep, and realized, as never before, the meaning of the above words; and when the first rays of the rising sun tinged the cloud-lands of the east, we found ourselves anchored in the Mersey River, between the cities of Birkenhead and Liverpool.

The first thing we noticed was the muddy

condition of the stream. Some, in derision, call the water of Toronto Bay soup, but we would call the water of the Mersey porridge. Dean Swift once said, on crossing the river and noticing the character of its waters, "The quality of Mersey is unstrained."

CHAPTER V.

LANDING AT LIVERPOOL—TRIP TO LONDON.

IN two or three hours after anchoring, a steam tender came alongside, and we bade farewell to the friends we had made during our eventful voyage of ten days. Just as we were about leaving the main deck, our friend, the sailor to whom we had made the suggestion some days before that he had better substitute the term hades for hell, as he was using the latter quite freely in the rigging when something went wrong, came along, and as we had become quite friendly, warmly shook hands, and said that as he had taken quite a fancy to us, he would immediately, on going ashore, drink our health in liberal supplies of good old Irish whiskey; and as he was an ardent young Irishman, his determination was significant. Of course, although we enjoyed his humour, yet our temperance principles forbade us expressing any approval, by word, look or sign, of his course. He informed the writer that his father was a prominent and wealthy man, and lived in Londonderry, and that he himself had got weary of the restraints of home and had run away to sea, and was now the black sheep of the flock.

The cry now rang out, " All aboard for Liverpool," and in a few moments we were steaming shoreward in the tender. Ten minutes later, and after other farewells, we trod the gangway and stepped ashore. It seemed singularly strange standing on solid ground once more, and we felt a little dizzy for some time. It seemed as though the ground was in motion, like the vessel.

The first tribulation after landing was to have our luggage examined by the Customs officials. As they proceeded in the discharge of their duty, we protested that it was unfair and an outrage to subject British subjects to such humiliating annoyance; but the leading officer in the department informed us that it was surprising what mean acts British subjects were capable of performing sometimes. When, however, they had completed the work of turning all our effects upside down, and discovered no infraction of the law, they politely said they were exceedingly sorry to have caused so much trouble and loss of time, and put everything so nicely that we left their department feeling that even a Custom House custodian could be a gentleman.

We now proceeded at once to the great Lime Street station, to take the ten o'clock train on the London and North-Western Railway for the world's metropolis. We had seen some very fine stations in Canada and the United States, but when Lime Street broke upon our startled and astonished vision, we at once gave it the palm for magnitude and architectural splendour. The building is one of the most magnificent of its kind in the civilized world. Of course, we must remember that all these great English railway stations have hotels attached, and a large part of the building is thus equipped and set apart for the accommodation of guests. Thus one explanation of the enormous dimensions of an English station is, that it is a combination affair.

But what solid-looking masonry! Like all the public edifices in the Old Land, these mighty fabrics seem to be built for the use not only of the present generation, but for the use also of generations still unborn.

On entering this mighty mass of masonry, we noticed trains of cars standing upon their respective tracks, and what struck us most forcibly was the vast difference between the cars of the Old Land and the New. These English railway vehicles are divided off into compartments, with seats upholstered principally in leather, a seat running completely around the compartment, except on the side of the door, and by paying a small extra fee one is allowed the privilege of having the entire space to himself, which perhaps on analysis may prove to be a little selfish.

These sections of cars, or perhaps more properly, divisions, are graded first, second, and third class. The entrance is always on the side of the car. We do not know whether any change has since been made in the scale of charges, but at the time of which we write, first-class fare was six cents a mile, second, four cents, and third, two cents. The third-class accommodation is almost as comfortable as the first.

The guards, or in Canadian phraseology, conductors, are fine-looking men, judging from their physical proportions. They are as carefully selected as our Toronto policemen, and wear a beautiful dark uniform, with shoulder strap and pouch attached, the latter very much resembling the cartridge pouch of a cavalry man. Their duties do not seem to be as numerous or onerous as Canadian or American conductors.

English trains are run with infinitely greater speed than trains on the American continent. The distance between Liverpool and London is about two hundred and eight miles, and those fast trains run it in about four hours, approximately averaging fifty-two miles an hour, including stoppages.

We thought we had acquired a great amount of strength to enable us to stand the long journey between Toronto and the Atlantic coast, but after the rapid journey through merry England, so rapid, indeed, that it almost caused nervous prostration, we did not think the strength amounted to very much, but this we learned that, whereas we could stand any amount of travel on our Canadian railways with little perceptible injury, we could not undertake a single journey of over one hundred miles in the Old Land without being completely done up. The explanation is the frightful speed on the one hand, and the leisurely mode of procedure on the other.

On the way to the capital we passed through a lovely country, beautiful farms, delightful shady groves, and elegant homes. England is, indeed, a veritable paradise, especially the part through which we now travelled. The almost indescribable beauty of its homes recalled the words of one peculiarly gifted:

> "The stately homes of England !
> How beautiful they stand
> Amid their tall ancestral trees,
> O'er all the pleasant land !"

We had often heard of the beauty of an English country lane, and accordingly anticipated

seeing something that would please the eye, but
the realization was infinitely more pleasing and
gratifying than the anticipation. To a thinking
man, the first thought that occurs, on seeing a
typical lane, in this land of historic interest, is,
what a desirable place for meditation, in the
restful quiet, and between the lordly elms, and
yew trees! What a place for the incarnation of
thought!

After enjoying to the full the beauty of field
and lane, we crossed the new Manchester ship
canal, a remarkable feat of modern civil engineering. So broad and deep was it that vessels of
great draught and heavy tonnage could pass
through. It was to be opened in a few days by
Her Majesty the Queen.

We also passed through Rugby, famous for
its school and also for being the home of the
celebrated Dr. Arnold, whose name is a household word on both sides of the Atlantic.

After leaving this well-known town, the country became still more beautiful; and although
we thought Ireland paradisiacal, we almost felt
sometimes as we journeyed on our way that

England was even more beautiful, but the aspect of the country changed as we passed through manufacturing and mining districts.

This cradle of the Anglo-Saxon race is particularly noted for its manufacturing industries, the products of which go out to all parts of the world, and her coal mines seem to be operated on an extensive scale. As we passed the mines we noticed many thousands of tons of coal lying near the shafts.

And now, after more than three thousand miles of travel by land and water, we were passing through the suburbs of London. They extend far out into the country; of course, this is a natural consequence because of its enormous size. Soon we arrived at Euston Station, and as we viewed the enormous proportions of this typical English hotel-station, the truth suddenly dawned upon us that we were in the world's metropolis, a city which, according to the testimony of those best able to judge, contained six millions of souls.

Many intelligent Englishmen assured the writer that the population of the vast city,

ENTRANCE TO EUSTON STATION.

including greater London, was not less than the above-mentioned number.

A word here with regard to the government of London will, we think, interest our readers. The city proper is not large, that is, the portion ruled by the Lord Mayor and council. We tried to ascertain, as nearly as possible, the correct population, but found it exceedingly difficult, as nobody seemed to know the exact number contained therein, but from what information we received would place it at one million five hundred thousand. If these figures are correct, then the population of greater London would be four million five hundred thousand persons, all ruled by local vestries, as the Londoners term them.

We went first to the hotel department of the great Euston Station and asked for their scale of charges. On scanning said scale, we found that they just charged so much for each particle of food, and, to prove that the charge was ample, stopping at this place would, with good rooms and first-class board, entail an expenditure of something like a sovereign a day.

In conversation with an American chemist

from Boston, who had been several times in England, some few days before arriving at the capital, he informed us that a man would be a fool to stop at a railway hotel, as they were so expensive. We felt a little annoyed at him for putting it so strongly at the time, as we purposed stopping at such a place, but the result of the perusal of the scale of charges convinced us that our American friend was correct, and as we had already played the fool too often during our lifetime, we determined to be wise for once and sought another stopping place. Applying at a nice family hotel not very far from the station, we found the charges a little more moderate and excellent accommodation—the table being served by a waiter in full dress with white necktie. We remained here for a day and night, very comfortably, and then sought a private boarding house near Russell Square, a very respectable quarter of the city. We applied at two or three places but they were filled with Americans; finally we reached a very nice house, with a notice to the effect that boarders were accommodated there, and upon application were informed

that a good room was just refitted and ready
for an occupant; so unhesitatingly, after finding everything in perfect order, and savoring
strongly of respectability, also that the lady of
the house was a Christian and member of the
Metropolitan Tabernacle, and had been a per-
sonal friend of the late Rev. C. H. Spurgeon,
whose portrait hung in life size in the sitting-
room; we took a room, with board by the
week, in this place, that seemed more like home
than any we had yet seen since leaving our
native land. Quite a number of persons boarded
in the house, which was much less expensive
than any first-class hotel, and, in point of style,
was little behind the best of them. There were
several servants in the house, and the table was
set in the most approved fashion and well
waited on.

We must confess that for a day or two at
first, we felt a little nervous and stilty, but that
gradually passed away as we became acquainted
with the inmates, and finally we felt perfectly
easy and at home.

The boarders were nearly all of English birth,

the men being all young and in business. One
was a member of the London Board of Trade,
and evidently a very clever fellow. We soon
learned, in conversation around the dinner table
and elsewhere, that these typical English people
were very intelligent, and as the writer walked
with some little dignity and enjoyed a good
dinner, he passed on more than one occasion for
an Englishman, and of course felt flattered.

A great many Canadians and Americans
have the impression that the English people in
their own land are stiff, formal and reserved,
but with regard to the latter qualities, at least,
we are prepared to say that we found them as
communicative and free and conversant as
Americans. This idea of haughty reserve has
been formed from the bearing of some persons
who have crossed the Atlantic and settled in
Canada or the United States, and who call themselves Englishmen, but from what we have seen
of those and from what we observed in the Old
Land, are prepared to say that these individuals
from whom some have formed adverse opinions,
are counterfeits, and not genuine, true Englishmen.

CHAPTER VI.

THE BRITISH MUSEUM.

THE first place of greatest interest that we visited was the world-famed British Museum. The building is of enormous size, and contains relics not only of an ancient civilization, but also of both ancient and modern barbarism. The departments of most interest to us were those containing relics of an early Grecian and Roman civilization, and we will venture to say that in those departments, at this time, there is more to be seen of Rome and Greece that will interest thoughtful people than in Rome and Greece themselves, as they exist in this closing decade of the nineteenth century.

Arms and weapons of all kinds, used long ago by the stern legionaries whose measured tread struck terror into surrounding nations, and whose trained hands have long ages ago crumbled into dust; swords, shields, breastplates, helmets,

spears, some of which doubtless did service on
Cannæ's fatal field ; others perhaps were borne
by stalwart sons of Rome, intoxicated with victory, as Hannibal's veteran warriors were borne
down the sweltering tides of ruin and annihila-
tion on Zama's bloody day ; some perchance
carried exultingly by valiant hands through the
burning streets of the Carthaginian metropolis;
some perhaps dropped from the nerveless grasp
of the last solitary defenders of one of the
mightiest empires that this world has ever seen.

A copper helmet we noticed particularly with
a part of one side eaten away by the destroying
hand of time, but otherwise in an excellent state
of preservation, that was claimed to have been
worn by an Etruscan soldier, before Romulus and
Remus, of fabled notoriety, founded Rome.

But apart from the arms and armour, and coins
and manufactured fabrics of those civilizations
which existed in the far past, we were especially
delighted and profited by the examination of
the extensive collection of ancient manuscripts.
Outside the departments above mentioned, those
of greatest interest, perhaps, were a number of

manuscripts containing a portion of the Scriptures. One, some of the psalms of David; another, a portion of the gospels; others, various portions of the revealed Word. The oldest of these MSS., which were written in the original tongue on parchment and vellum in uncials and capitals, dated back to the fifth century of the Christian era, and is known as the Codex Alexandrinus, which contains the entire Old Testament, and also the New with but few omissions. Some dated from the seventh century, and others from the tenth, twelfth and thirteenth, etc.

We have thus endeavoured to give some idea of what this wonderful structure contains. Of course, it is impossible in these pages to give a detailed account of everything that this wonder of creation contains; it would not only take an entire volume, but many of them, to give anything approaching to an adequate conception of the multitudinous collections and relics contained within those mighty walls, so that any attempt in this direction would be presumption on our part.

We had read a great deal during the past ten
years, and studied some, and thought we knew
considerable, but when we entered the department of the Museum containing the books that
had been accumulating for ages, and noted their
number, so vast that we felt impelled to use the
Scripture language, "a multitude that no man
can number, their name is legion, for they are
many," we felt that we had but picked up the
first pebble from off the ocean shore of knowledge. If any of our readers have the slightest
tendency to unduly magnify their knowledge,
we would recommend that they cross the "big
pond" that separates the Old and New Worlds,
and take a walk before breakfast some morning
in that department of the British Museum which
contains the bound volumes of many circling
centuries, one hundred thousand in number, and
we venture to say that, as they confront this
heterogeneous mass of dead men's brains, they
will suddenly awake to the fact they have not
yet learned how little they actually know.

In leaving the precincts of this time-honoured
building, with all its associations and memories,

we could not but feel that never before in our lives had the philosophical laws of association and suggestion worked to better advantage than in studying those things, hoary with age and closely related to the history of the world.

CHAPTER VII.

THE BANK OF ENGLAND.

THE next place of interest visited was the Bank of England, the greatest monetary institution on earth at the present time. It was founded in the year 1694, by Mr. William Paterson, a Scotchman, with a capital of £1,200,000 sterling. The subscribers of the capital received 8 per cent. interest and £4,000 a year as expense of management, which is committed by charter to a governor and twenty-four directors.

The Bank has, during its history, passed through a great many vicissitudes of fortune, being compelled two years after its establishment to suspend payment, and but for the assistance of the Government would have utterly collapsed. To guard against further difficulties of this kind, the capital was increased from the above figures to £2,201,171 sterling. The institution passed through another crisis in the year 1745, when the Highlanders, under the Young Pretender, invaded England, and caused a run on the Bank, which nearly ruined it; but the retreat of the invaders finally saved the establishment. In the closing years of this eighteenth century, what with loans to the German Government and also heavy loans to the Home Government, and rumours of invasion, which caused another run on the Bank, all the available cash and bullion in its coffers amounted to but £1,272,000. One day more, it was clearly foreseen, would ruin the concern, and accordingly a clever ruse was resorted to.

An order-in-council was immediately passed, by which no more cash was to be paid out until Parliament took action in the matter, and of course by the time any legislation was enacted the danger had passed. The capital of the institution at the beginning of the present century was about £14,000,000, with about £4,000,000 in the form of bullion. Its history throughout the present century has been comparatively uneventful, and the present capital is in the neighbourhood of £15,000,000 sterling, with about £5,000,000 or £6,000,000 of it in gold, silver coin and bullion.

In visiting the Bank, we had the pleasure of seeing an enormous quantity of specie—canvas bags filled with gold and laid two alongside of each other, and so on alternately, until they reached several feet in height, and silver piled like grain in bins.

Where such quantities of coin are handled they do not count, but weigh. One pound troy weight of silver is equivalent to $10, and the same weight of gold to $240. In a financial establishment like this great precautions must necessarily be taken to guard against loss by theft or raid. Forgery is difficult to guard against, as the history of the Bank clearly proves, it having lost on one occasion by this means over £300,000. But to guard against the former dangers, every night a company of regular troops of the line mount guard, and, as an extra precautionary measure, during the hours of darkness the vaults are flooded with water. We are not surprised at

BANK OF ENGLAND.

these thoroughly efficient measures being taken when we consider the enormous wealth contained within the four walls of this building, which is low in structure, but strongly built, resembling a fortress more than a bank. In all the corridors, and, indeed, in every department, we noticed men in gorgeous livery pacing back and forth, and it struck us that they were not here solely for ornament, but, if the need existed, would stubbornly defend England's wealth.

Within this wonderful structure is a machine, regulated with great precision, through which sovereigns are passed on being received. All that are of light weight are thrown to one side, whereas those of full weight are passed through as approved. Those that are thrown out are to be recoined, and the approved that are passed through are transferred to the vaults of the Bank.

It is scarcely necessary to say that the Bank of England is the first institution of its kind on earth. It is said that the Bank of France comes next; and, all honour to our native land,

5

the Bank of Montreal is claimed to be the third
greatest financial institution in the world, with
a capital of $12,000,000—$6,000,000, or one-
half of the entire capital, lying in the vaults
as a rest fund.

One reason for the more than ordinary promi-
nence of the Bank of England, apart from its
enormous capital, is the fact that the Government transacts all its business in connection
therewith. It is conceded that there were
times in its history when the continuance of
its charter was doubtful, and would probably
have been discontinued but for the readiness
of its management to assist the Government
in time of need by large loans, and because
of this generosity on the part of the governor
and board of officers, the Crown officials spare
no pains to preserve the status of the establishment, and also readily and cheerfully grant
the use of armed troops for its protection.

CHAPTER VIII.

GUILDHALL—HIGH HOLBORN STREET—THE STRAND— UNDERGROUND RAILWAY.

AFTER visiting this renowned institution, hoary
with the memories of more than two hundred
years of public service, during which it not only
passed through periods of great prosperity, but
also through its waters, dark and deep, of afflic-
tion, and through years fraught with trial and
danger, we dropped into the Guildhall, a public
edifice of note and importance, which may, properly speaking, be regarded as the Town or City
Hall.

The Court of Common Council, the Court of
Aldermen, the Chamberlain's Court, and a Police
Court, presided over by one of the aldermen,
make this historic edifice their place of assembly.
The original Guildhall was erected in the year
1411, but was totally destroyed by fire in the
great calamity of 1666. In 1789, it was rebuilt
in its present form. The main hall is 153 feet

in length, 48 feet in breadth, and 55 feet in
height.

Thus for nearly five centuries, since its
original founding, has this place become noted
for the profuse magnificence of its civic feasts.
One of the greatest ever held within the
masonry of this time-honoured edifice was in
the year 1500, when Sir John Shaw, a wealthy
goldsmith, who had been knighted on Bosworth
Field, became Lord Mayor of London.

The main hall, the dimensions of which are
given above, is bordered with costly statuary.
Busts of former Lord Mayors, and many of
England's great military and naval leaders,
abound. Chief among them we noticed that
of the Duke of Wellington, Lord Nelson, General Charles E. Gordon; and of statesmen, the
Earl of Chatham, the younger Pitt, and others
of lesser note.

Here again, as in the Bank of England, were
men in the most gorgeous livery, giving some
faint idea of the pomp, and pageantry, and cir-
cumstance that enshrouds the governing power
of this great city.

THE NEW LAW COURTS.

An Art Exhibition was being held at the
Guildhall at the time of our visit, and we cannot speak too highly of the quality of these
productions of the old masters. Though not an
artist, and making no pretensions to the posses-
sion of sufficient ability to criticise art, yet the
most untutored along this line could not fail
to see the infinite superiority of the work of
the distinguished artists represented here. And
although our Canadian Schools of Art are excellent, and we believe fully up-to-date, yet we
would advise their graduates, if possible, to
cross the sea and study the works of illustrious
personages in their profession in the Old World.

The paintings exhibited here were numerous
and varied—battle-pieces, scenery, portraits, etc.
In the list of the former were some of England's
greatest battles by land and sea, and so intensely realistic were they, that they only
lacked the roar of cannon and rattle of mus-
ketry and shouts of the combatants to make
the student feel that he was actually in the
midst of the conflict. Feeling thus, reminded
us of work done by a great artist in the form

of fruit fully matured, and birds entering the
room where the work was displayed, pecked at
the cherries, mistaking them for natural fruit.
And note the fact, that the artist who can make
the observer of his production feel that it is a
reality, has attained to the highest perfection
in art.

The sceneries also seemed as strikingly natural
and realistic as the battle-pieces, many of them
being of exquisite beauty. This can be said
with as much certitude of the paintings representing flowers and animals; their merit can be
seen in their striking realism.

On our way to Russell Square, after visiting
Guildhall, or town hall, or hall for a collection
of persons gathered together for mutual benefit,
representing a guild, we passed through High
Holborn Street, one of the leading thoroughfares
of the capital. The amount of traffic on this
great metropolitan highway is astounding; it is
almost impossible to cross the street in safety,
especially so to one who was accustomed to
small cities where the traffic was not nearly so
heavy. The first few times in which the writer

crossed, he made a very sorry spectacle, dodging around among the vehicles, which literally swarmed upon the thoroughfare. At last we took the attention of a policeman, who seemed greatly amused at our efforts to cross, and he very kindly and courteously showed us a safer plan of procedure.

On visiting the Strand a little later, having heard that it was especially noted for its enormous traffic, we found the same swarm of vehicles as on Holborn Street, and the same difficulty in crossing was experienced. To show that we do not unduly magnify the difficulty or danger of crossing these leading streets, it was told us by an intelligent Londoner that every week there are seven hundred persons injured on street crossings and conveyed to the various hospitals of the city. There are no electric cars in London. This is not accounted for by setting forth the theory that the city is behind the times. Such is not the case, but because of the thronging of its leading highways, it would be absolutely impossible to safely use electricity as a motive power. We have proof of this in the fact that many smaller and less important cities in England have electric cars.

The streets in the capital are substantially paved. We noticed in some places that the common hard-head was used, and because of this the surface is a little uneven, and vehicles make a considerable noise, and do not wear as long as on smoother pavements. Of course, many of the streets are otherwise paved.

After having tried many different kinds of pavement, doubtless London has waked up to the fact, like many other cities, that in order to insure durability, rougher materials must be used, regardless of wear and tear of rigs and of noise. The English idea of the substantial is so matured and acute that it will not, even on the streets, allow the use of anything of a flimsy character. The public buildings, indeed all the buildings and productions of engineering skill, give evidence of substantiality almost beyond conception. Along many of the streets, especially in the more respectable portions of the city, the houses run four stories and a basement, thus giving a fine impression and adding very

HOLBORN VIADUCT, BRIDGE ACROSS
FARRINGDON STREET.

materially to the appearance of this place of renown.

For the convenience of the public in the absence of the electric motors, 'busses are run by

ROYAL EXCHANGE.

horse-power, and are not only fitted with seats internally, but also on top, and it is a familiar sight in the Strand and in Holborn, New Oxford, and other leading streets, to see the two rows of seats upon the top of each 'bus

filled with passengers, the silk hats of the gentlemen shining in the sunlight, conspicuous everywhere, as they are more commonly worn in the
Old World than in the New.

In travelling through the city, we frequently
took the top of the 'bus, and grew quite fond of
riding in the open air, as the weather was mild
and pleasant. In almost every square, we noticed
an inclosed space, and on investigating the
matter, found that within said enclosure was an
opening to furnish light for the use of the underground railway.

It would be unpardonable in these pages not
to say something about this wonder of modern
times, a city of six millions of inhabitants completely undermined by a network of tunnels
through which trains are run every few minutes.
To find out exactly how these railways were
worked, we descended the staircase of chiselled
stone at one of the stations, and found it, quite
contrary to expectation, well lighted and a really
pleasant place. A number of persons were sitting on benches in front of the station, waiting
for a train. The building, though small, was

neat and comfortable, and specially adapted for an underground station.

The cars were very much like ordinary English railway carriages, but much smaller. A good story was told us here of a young cockney, who, by the way, was a little fast, and desiring to go

SCHOOL FOR THE INDIGENT BLIND.

to a distant part of the city, took an underground railway train, and after taking his seat, noticed a very handsome young lady on the opposite side with her attendant, a coloured girl, beside her, who, by the way, was as homely in appearance as her young mistress was beautiful. The fast

young man decided that as soon as the train left
the station, he would purloin a kiss. Soon it
started, and all became as dark as Egypt, when
the young man, calculating the precise position
of the young lady in the darkness of the tunnel,
rose up, leaned forward and secured the coveted
smack. A moment later the train had arrived at
its next stopping place, and all again became
light, but to his chagrin and horror the young
swell discovered the astounding truth, that the
young lady in question had quietly changed
places with her coloured attendant, and that he,
unaware of such change, had embraced the
negress.

CHAPTER IX.

CRAVEN CHAPEL—ST. JAMES' HALL—FIRST SUNDAY IN
LONDON.

ON leaving the Metropolitan underground railway station, we proceeded to our lodgings, to
prepare for the holy Sabbath, as next day was
Sunday. On the morning of this our first Lord's
Day in the outgrowth of ancient Camelodunum,
we were, of course, particularly anxious to hear
some of its most distinguished preachers. One
of the boarders at our lodging place, an elderly
Christian lady, offered to pilot us through the
streets to Craven Chapel, where Mr. Hughes was
announced to officiate in the morning; so we
gladly consented, and reached the chapel at the
appointed hour, with our aged guide. The
church was a quaint-looking old place, having
been a Congregational place of worship previous
to the time in which the present distinguished
preacher took charge. The congregation was
large, for a morning service, and the people

6

looked as though they belonged to a very respectable class of society. The pastor, Rev. Hugh Price Hughes, was a man of about average height, and of intellectual appearance and pleasing manners. Throughout his entire sermon he endeavoured to impress upon his hearers the necessity of a full surrender to Christ, and the need of earnest prayer. In teaching his people the necessity of whole-souled consecration, and of the use of the medium prayer, and how it should be applied, he made a statement that we can never forget, saying that some people live too long, and counselling his hearers not to misapply prayer by asking God for long life, when, perhaps, if the petition were granted, it might be the worst thing that ever happened them, as by living a little too long they might spoil all the good they had ever done. He also went on to say that modern education was too often antagonistic to heart consecration, and that the most dangerous men in all England were the educated classes, many of the graduates of world-renowned Oxford going out into society to pervert Christianity, instead of endeavouring to strengthen her hands; and whereas, they might have had their minds marvellously expanded, and been a benediction to their fellowmen, they have had their understanding darkened, and have not even learned the true meaning of the motto of their *Alma Mater*, " Let God enlighten thee."

Before closing this wonderfully practical sermon, however, Mr. Hughes showed himself a friend to all true education, as we would expect from one who was himself educated. After preaching, a baptismal service was held, which was very impressive; and after all was over, our lady friend, who, by the way, was well acquainted with the celebrated preacher, with whom she often corresponded in connection with Christian work, presented the writer to him, who was very courteously received, and invited to attend the service at St. James' Hall at seven o'clock in the evening. We could not refuse the kind invitation, and, accordingly, went alone; and, on entering the main ante-room, whom should we meet but Mr. Hughes himself, and he at once invited us to take a seat on the platform, but we

protested and said that we had rather take a
seat in the gallery, where we would not be
noticed, as we were dressed for travelling and
not for show; but he would not take "no" for an
answer, and we were compelled to face the
inevitable as cheerfully as possible, and sit for
nearly two mortal hours before several thousand
persons, as the great hall was crowded to suffocation. The sermon on this occasion was a
masterpiece of sacred oratory; and as the vast
assemblage was dotted here and there with the
scarlet tunics of some of England's warlike sons,
the preacher of the evening turned with great
earnestness and power to one part of the gallery
where the greatest number of soldiers sat, and
said: "You men, whose business is war, we
counsel you to-night to turn to a loving Saviour
and accept the Gospel of peace as your guiding
star, as you journey toward the shores of
eternity." At the close, an invitation was given
to all who desired to become Christians to stand
up, and we judge that at least fifty persons rose
to their feet, and, to our infinite joy, a number of
them were from among the red-coated soldiery
of Britain.

CONGREGATIONAL MEMORIAL HALL.

At the close of the service, Mr. Hughes still continued his courtesy by introducing the writer to several of the leading officials of St. James, and asked us to go to the inquiry room and speak to some of the seekers. We did so, and, to our joy, two young men to whom we spoke declared that they were there and then ushered into the light of the better life, which resteth at noon. One of them confessed that he had led a very licentious life. Thus God makes use of us, if we are willing to be used, in any clime or under any conditions.

On going to and from these services we noticed that the streets of the "World's Hub" are very quiet on Sunday ; indeed, this place, with its millions of inhabitants, is almost as quiet as our model Canadian city, Toronto, on the Lord's Day, but in all candour we must acknowledge that for six days out of seven, it makes ample atonement for this quiet.

CHAPTER X.

The Tower of London.

OUR attention was next drawn to a place of more historic interest, perhaps, than most places in the Old Land, memorable as being the scene of events closely connected with the most racy pages of English history, in some respects also of tragic memory, namely, the Tower of London, which was in feudal days a powerful fortress, then degenerating into a state prison of ghastly association. This place is an irregular quadrilateral collection of buildings, situated on slightly rising ground, on the banks of the lordly Thames in the eastern part of the city. The space occupied by the Tower buildings is almost thirteen acres, surrounded by a shallow moat, or ditch, which is usually dry, but which could, in a few moments' notice, be filled with water by the garrison. The moat is bordered on the inside by a high castellated wall, broken here and there by huge flanking towers; within this first wall is another of still greater height, and within this are the various barracks and armouries, and in the centre of all these the stately and historic White Tower, which was built by Gundulph, Bishop of Rochester, in the time of William the Conqueror. It is the great centre of interest in connection with the Tower buildings, and its walls are sixteen feet in thickness of solid masonry. This ancient piece of architecture was the court of the Plantagenets.

With regard to the Tower generally, some writers maintain that it was built by Julius Cæsar, as a Roman fortress, but there is no written evidence to prove that the White Tower was built before 1078. The Saxons seem to have had some earlier structure there, as is proven by the massive foundations discovered in subsequent times, but of this fact we find no proof in history.

During the reign of the first two Norman kings, the Tower was wholly used as a fortress. In the time of Henry I. it existed as a state prison. During the reign of this monarch it was gradually strengthened, and also by his

successors, until at last it became one of the
greatest strongholds of the time. The kings
frequently resided and held their most brilliant
courts at the Tower, and had sometimes to with-
stand sieges and blockades from rebellious
citizens.

A great many executions, for political offences,
or supposed offences, took place here, and in one
portion of the Tower are ghastly reminders of
such scenes—the axe, headsman's block and
mask. What strange thoughts crowd the brain
as one looks upon the very block upon which
the beautiful and accomplished Lady Jane Grey,
the Earl of Essex, Sir Walter Raleigh, and at
a later day, Lord Kilmarnock, Balmerino and
Lord Lovat, after the rebellion of 1745, laid
their devoted heads, and the murderous-looking
axe that lopped them off, and the mask worn by
the grim headsman as he performed his bloody
task; but happily, with the establishment of a
limited monarchy, those bloody scenes enacted
in the drama of the Tower's history ceased, we
trust, forever.

Very touching and pathetic are the inscriptions cut upon the walls by the hapless individuals confined within those ancient precincts —among them, the seven bishops, of whose confinement our readers will remember having read in their English histories, and later of Wilkes, Horne, Tooke, and others. In 1841, the Tower had to battle with the fire fiend, which was communicated to the armouries and resulted in the destruction of many valuable relics of the past, and of several thousand stand of arms.

The government of the Tower is vested in a constable, usually a military officer of high rank, who with a deputy under him, also an officer of rank, has charge of the corps of yeomen—or, more familiarly, beef-eaters—on duty continually there. Also, during our visit to this place, made famous by a thousand memories, the 9th Regiment of Royal Rifles occupied the barracks. We conversed with many of the soldiers, and found them, as a rule, intelligent young fellows. They wore the scarlet tunic and spiked helmet, and were armed with the Lee-Metford magazine rifle, and looked quite soldierly.

In one of the Tower buildings are kept the

arms and armour of the belted knights of old, burnished and bright as if manufactured yesterday. We had heard that men were taller in the nineteenth century than the men of any other age, and one of the proofs given was the armour used in the Middle Ages and preserved in this remarkable place. The theory set forth was this, that said armour was so small that fair-sized men of our day could not encase themselves within it. We took some pains, however, to ascertain if this was correct, and found to the contrary, that few men less than six feet tall ever buckled on a single suit of this armour. The weight of these suits was surprising; one, for example, that had been used about four hundred years ago, weighed one hundred pounds avoirdupois; another, worn by one of the Henrys, tipped the scale at ninety-six pounds.

The thought enters the mind at this stage, would it not be impossible for men to march in such weighty armour? It certainly would for any considerable distance; but we must remember that the men encased therein were invariably mounted, many of them on armour-cased horses.

As proof of this, we noticed some splendid suits of horse armour of enormous weight and bright as silver, so that it would be next to impossible to find an aperture through which an arrow could be sent, at any distance, in battle.

In this department, also, were battle-axes, manufactured very much on the principle of the hewer's broadaxe, only much lighter, and some smaller, with handles six or seven feet long, used also in the Middle Ages ; and, what amused the writer above everything else, were some old muskets used in the time of Queen Anne, with enormous bore and hammer as large as the hammers of three ordinary modern percussion locks, and most amusing of all, this ponderous weapon, which was too heavy for a soldier to aim off-hand, had a contrivance attached to the barrel very much resembling a pitchfork, with the tines fastened one on each side the weapon and running together underneath, in the lower part of which was a socket that fitted upon a stick driven in the ground, and with this rest the soldier was enabled to aim the prodigious rifle, which, judging from the size of the bore, would take a whole handful of powder for a charge.

There were also in this place guns, horse-pistols, bayonets, swords, drums and colours taken from the victorious field of Blenheim, reminders of English valour on continental soil.

In an adjoining department, which was not open to the public, we could see, through glass, long rows of rifles of modern workmanship and of doubtless deadly efficiency, closely set in frames in an upright position. With regard to number, they were so numerous that it was utterly impossible to either count them or guess the proper figures; but as the Government now uses the Tower for a military storehouse and arsenal, which is completely under the control of the War Department, we may truly surmise there are numbers sufficient to arm and equip a very large army.

But of the courtyard of the Tower we have, as yet, said nothing. It contains the greatest collection of ordnance of all descriptions, per-haps, in the world at the present time—cannon and mortars taken from the French and Spanish

in Wellington's victorious campaigns in the
Peninsular war; also others, relics of the
Crimean war, of the victorious and bloody
battles of Inkerman, where the British soldiers
greatly distinguished themselves for their reck-
less bravery, and Balaklava, memorable for the
charge of the Light Brigade, the noble Six
Hundred; also cannon captured on the memorable day when the Alma's heights were won,
when the Russian hordes were hurled back in
the gloom of defeat, losing one-third of their
number. We noticed mortars in this heterogeneous mass of the enginery of war that threw
stone shells, we believe, two feet in diameter.
We tried to lift one, it being hollow inside, of
course, and the shell appearing quite thin, but
found it impossible. It would take at least two
full-grown men to place it in a mortar when
loaded.

We could not help thinking that a visit to
this place, reeking with historic memories,
would be extremely distasteful to Frenchmen,
Russians and others, because of the presence of
these trophies of their discomfiture. In one of

the departments is a section of mast of Nelson's flagship, the *Victory*, that was pierced by a solid shot at the battle of the Nile. The hole where the shot pierced the mast was not ragged as we would have expected, but clean cut. This is invaluable as a memento of England's being saved by victory on the shores of Egypt, under the leadership of her greatest naval hero, Admiral Nelson.

Also, there is a fragment of the keel of the *Royal George*, and one of her guns, the flagship of the British navy that capsized off Spithead in the last century, with the admiral and nine hundred of England's picked men. All were lost. The cause of the accident was the tilting of the ship to make some repairs, as the lower part of one of her sides had been damaged in some way; and as she carried 120 guns, sixty on each side, in three tiers, although but slightly heeled over, it caused the sixty guns on one side to break their fastenings and run across the deck to the opposite side, capsizing the vessel instantly, all being done so quickly that the most capable seamen were utterly helpless. Not a moment's warning was given, but almost in a twinkling the vessel turned bottom up. All England was plunged into grief over the terrible calamity that robbed her of an admiral and nearly a thousand of her best trained seamen.

We examined the gun that had been fished up, after lying many years under the briny waters of the sea, the saline quality of which had eaten into the barrel of this engine of death, that had so long and harmlessly lain in its watery tomb.

While running along this line, we might also say that the entrance to the Tower is guarded by a beautifully chased piece of ordnance, with a barrel from sixteen to twenty feet in length, mounted on a splendid carriage, presented by an East India Maharajah to Her Majesty Queen Victoria.

And now, last in connection with London's world-famed Tower, we come to the crown jewels of England, preserved within large glass cases. In one of the many departments of this wonderful place there are crowns, coronets, sceptres, staffs of office, etc., all guarded by the yeomen or buffeters, or in common, vulgar parlance, beefeaters. We inquired of one of these if the jewels were of pure gold, and he informed us that they were all twenty-four carats fine. We made inquiry also concerning their value, but in no case did it seem to be known. They are doubtless of almost priceless value, judging from their size and the weight of pure metal without alloy, valued at two hundred and fifty dollars to the pound, troy weight, independent of precious stones which abound. We are not surprised at the ignorance manifested along this line; as although it would be comparatively easy to compute the value of the gold, yet with regard to the precious stones, it would be almost impossible. Suffice it to say, that Her Majesty's crown alone, which is so small that it could readily be placed upon the open hand, and which is worn only on State occasions, is worth one million pounds sterling, or at least claimed to be worth that.

CHAPTER XI.

REGENT'S PARK—HYDE PARK—THAMES EMBANKMENT.

AFTER our exhaustive and painstaking visit to those scenes so pregnant with historic interest, we turned our attention toward the restful parks, first visiting Regents, one of the most beautiful in the world, with delightful walks and drives, fountains and miniature lakes; in a word, everything that could please the fancy of the most fastidious, even to flowers, in beds laid out with exquisite taste. This park is the favourite resort of the common people, and the best of order is preserved, a number of guards being employed and wearing a neat and beautiful livery.

An amusing incident occurred during our visit to this place. A woman had been sitting on one of the benches just inside the main entrance to the park, when suddenly she rolled over on the seat and in a few moments straight-

ened up again, and began working as though in
convulsions. Seeing the dilemma, and unwill-
ing to approach a strange individual of that
particular sex, we at once called the attention
of one of the guardsmen to the fact that help
might be needed, but he laughed good-naturedly
and said that she was just a little boozy, after
being on the spree the night before; and lest
we should think that she was an Englishwoman,
the guard said that she was "a bloomin' German,
anyhow, as nobody else would go on that way."
And all through our sojourn in England, we
noticed the marked absence of open vice which,
as a matter of course, abounded in the low
quarters of London and other large cities; but
as the writer was not familiar with the haunts of
vice in these places, can only speak of what he
actually observed, and on relating our impressions to an Englishman whom we knew to be
familiar with the habits of his countrymen,
were informed by him that the English people
did not, as a rule, sin openly when they chose
to sin, or flaunt their vice before the world
when they saw fit to be vicious, but all or

nearly all was quietly done behind the scenes, and as the cool-headed guardsman in Regents Park would have us believe, was nearly all done by foreigners from the continent of Europe.

But we now drop the description of this paradise of the poorer classes for a description of Hyde Park, where fashionable London disports itself. In this place is Rotten Row, where the fashionable and aristocratic classes practise their fine blooded horses. Both ladies and gentlemen were riding here during our visit, and, although having little taste for jockeying, and being but a poor judge of horses, we could not close our eyes to the fact that some of the best horses in the world were here. Some superb animals of the hunter breed were being exercised by exceedingly skilful riders of both sexes. We must not pass on without saying something concerning the place which bears this horrible name. It is paved to a considerable depth with some spongy wood, somewhat resembling cork and of a colour very similar. The reason for this yielding pavement is, that if one of the riders was thrown, there is comparatively little danger of fracturing bones, as it would be very much like falling on sawdust; hence the name, derived from the yielding or spongy or decayed nature of the pavement, Rotten Row. Passing from this place of disportment for mounted pleasure-seekers, we found ourself on the leading drive, and, as we were anxious to see some of the best equipages, took a stand right alongside. Some splendid carriages, to which were attached well-bred horses, passed, their occupants being aristocratic-looking personages.

At last an unusually fine turn-out passed. There were two prancing teams attached to a handsome carriage, with reserve whiffletrees and neckyoke hanging upon hooks behind. We felt sure that it was occupied by very prominent persons, and inquired of a policeman as to their identity. He informed us that the turn-out belonged to General Sir Henry Hewitt, the wealthiest man in England, or more properly, one of the wealthiest men in the kingdom.

It scarcely requires close observation to pick out those of aristocratic position and lineage among the frequenters of this notable park. There is a nameless something in their turn-out,

MARBLE ARCH, HYDE PARK.

mien, dress or general appearance which tells of
noble birth and prominent social status; and as
we walk around this delightful spot and notice
the beautiful fountains in full play, the drives
through majestic shady trees, and the large
fancy-shaped flower-beds, with their numerous
and lovely varieties, and stately monument of
England's greatest son, do not wonder that this
is the centre of attraction for the beauty and
chivalry of this metropolitan city.

After visiting this land of Beulah, we started
for the Thames Embankment, and after half an
hour or more of pushing through crowded
streets and dodging multitudinous vehicles, we
reached the Embankment and found it splendidly
constructed—masonry very similar to that of
the docks at Liverpool.

All along the river's bank was bordered with a
row of fine healthful trees, evidently well cared
for, and inside this again were other rows,
and a splendid drive between. We could not
but notice the beauty and vigour of the English trees everywhere, except in Liverpool, and
for some reason we could not ascertain, they
seemed stunted failures in that city.

The bridges that spanned the Thames with their great circular arches, were so long that one arch of some of them spanned the Embankment, all carriages passing underneath, making the drive quite novel.

On the bank, using the abbreviated term, stands the celebrated needle of Cleopatra, brought to England with the greatest difficulty, and placed in an upright position, only a few yards from the river. It bears inscriptions in the hieroglyphic language of Egypt, the key to the deciphering of which was found in the discovery of the Rossetta stone by the *savants* of the French expedition. We cannot say of the needle as we have said of many things which we have seen in our travels, that it is beautiful, but if we cannot truthfully apply this term to it, we can at least say it is strange and weird-looking, and to thoughtful minds suggests many things in connection with that land of wonder from which it came, with its civilization thousands of years old. Singular that in an unhandsome shaft of granite, with its pointed top and strange, mysterious characters, we should read the history of ages, with their teeming associations,

CHELSEA EMBANKMENT.

CHAPTER XII.

ST. PAUL'S CATHEDRAL.

IN following the river, to acquire a more thorough knowledge of the shipping, and to see more of the Embankment, we were led almost to St. Paul's Cathedral, and finding this out, we at once proceeded to visit the place. Our readers will remember the old saying, "it is better to be born lucky than rich," and certainly if never fortunate before, we were exceptionally so on this occasion, for we just arrived in time for a musical service, and sat for nearly an hour listening to the sweet strains of the grand old organ, accompanied by a large number of trained voices. A great many persons were present, all eagerly drinking in the golden melodies. After listening until a good many of the hearers had left the grey old cathedral, and the musical programme was about ended, we walked around the interior, where were monuments to the memory of many of

England's dead warriors. Conspicuous among them was one to the memory of the Iron Duke, the victor in a hundred fights.

Whatever faults England may have, one thing is noticeable to those who cross the ocean and visit her public edifices. She has not the fault of forgetting those who, in the hour of her need, have stood by the old flag, and as we noticed this, in the many costly monuments she has erected to perpetuate the memory of her sons, who have done valiantly, were not surprised at the recent utterance of an American senator, who said that England, on one occasion in this present century, hearing of the unlawful and unjust imprisonment of one of her subjects in a remote part of the earth, among semi-barbarous people, organized an expedition, and at a cost of twenty-five millions of pounds, rescued the solitary imprisoned subject. When a nation would do that for one out of a population of teeming millions, is it any wonder that her sons are so filled with patriotic love, that they would cheerfully give up their lives for her on any field ?

We may say this of the monumental piles,

erected to commemorate the valour or devotion
of her humblest devotees—she is so careful to
honour faithfulness in any—need it surprise us
that, seeing this, millions would as cheerfully go
at her command to their death as to their bridal?

Another splendid piece of sculpture with
reclining figure resting on couch and pedestal of
copper, commemorating the bravery, Christian
integrity and fortitude of General Sir Charles
E. Gordon, erected by his brother in loving
remembrance, bears upon its side this inscription in letters of gold : " He gave his goods to
feed the poor, his courage to defend the weak,
and his soul to God."

On one of the walls of the time-honoured building is an inscription, and sculptures of military
figures with bowed heads and reversed arms, in
memory of the veteran officers and soldiers of
the Middlesex Regiment, who fell in battle
during the Crimean war.

The general interior of the cathedral, apart
from monuments and inscriptions, is gorgeous in
the extreme, the decorations being surpassingly
beautiful.

CHAPTER XIII.

KENSINGTON MUSEUM—NATURAL HISTORY MUSEUM—
UNIVERSITY OF LONDON.

OUR next day's sight-seeing was in Kensington
Museum, which in many ways may be said to
be an annex of the British Museum. In connection with it are beautiful gardens containing
the Albert Memorial, a fitting tribute to the
memory of a wise and virtuous prince. The
building itself is almost as extensive as the
parent museum mentioned above, and contains
literature of almost priceless value, because of
its antiquity, also collections of arms and armour
of ancient times, as well as of the Middle Ages.
One old sword that we were specially interested
in, with its blade half eaten away with rust,
and made on the principle of a huge knife, was
dug up in Kent, and proven to have been used
in the times of the early Saxons.

There was also a vast collection of coins, some
very valuable and of great beauty, others of

ALBERT MEMORIAL, KENSINGTON GARDENS.

8

ancient date. The most beautiful of all the coins, in our estimation, was the Portuguese doubloon, a gold piece, considerably larger than the American twenty-dollar piece, and worth about twenty-six dollars.

Conspicuous among the valuables was a sword of finest temper, inlaid with gold and studded with precious stones, worth, doubtless, thousands of pounds, that had been taken in the conquest of India from a Maharajah.

A large number of medals were also displayed of more or less value.

In another building connected with the Museum, we had the pleasure of seeing the first sewing machine ever manufactured. It was all of wood, and so crude in comparison with the superb machines of modern workmanship, that it furnished infinite amusement to the writer.

In the same building, though in another department, were models of England's old wooden walls, the battle ships that won her most glorious victories, in the days of the heroic Nelson, and also of some of her modern steel-clad cruisers and battle ships of the line. One model we spent a long time over in sadness, namely, that of the *Victoria*, sunken in collision with the *Camperdown*, in which England once again lost several hundred of her choicest sea-men. These models give an excellent idea of the efficiency and power and symmetry of men-of-war of all classes.

The collection in the Natural History Museum is also rare and interesting—all kinds of ani-mals, birds, reptiles, insects; in a word, every-thing from an animalcule to a whale.

There were exhibited several chemical proper-ties, all of them found in the human anatomy, such as phosphate of sodium, phosphate of potas-sium, chloride of sodium, carbonate of calcium, etc. These are also found in the soil of the earth, proving the truth of the Bible story of the creation of man, when the Lord made him out of the dust of the earth, and breathed into his nostrils the breath of life, and he became a living soul.

Thus at a period when the so-called Higher Criticism threatened to explode the teaching of the book of Genesis, the noble profession of medicine came to the rescue with these astounding facts, and incontrovertibly clinched the truths of inspiration.

The proof, then, is the existence of the same properties in the soil and in the body of man.

In this building is a life-size sculpture of Charles Darwin, not erected to his memory or placed there because of his being a materialistic evolutionist, but because of his having attained to distinction as a naturalist and specialist.

One thing noticeable in the natural history collection is, many of the animals have been taken from our great Canadian Dominion, a thought gratifying to our countrymen everywhere. The place is very extensive where the different species abound, the building a magnificent structure, admirably adapted to the purpose for which it is used, and the student will find within its walls information along this particular line that will broaden his perceptions for the remainder of his natural life—buffaloes, gnus, bears, wolves, camels, leopards, tigers, panthers, lions; and in the reptile department, the boa-constrictor, cobra-de-capello, rattlesnake, and a variety in all lines.

In the bird department, almost everything, from the ostrich that roams over African sands, to the little humming-bird that sips the aroma of northern flowers, is to be found.

The information received amid such surroundings will not soon be forgotten, and we felt that this visit, to others of England's places of chief interest, was an inspiration indeed.

We finally closed a memorable week by a visit to University College, one of the greatest seats of learning in the Old Land, noted especially because of its being a chartered institution with university powers in the greatest city in the world. Men come from all parts of the earth to take post-graduate courses in this place, with its splendid equipment and degree-conferring power. Through the courtesy of those in charge, the writer was privileged to visit the different departments, which were of great interest, especially the departments that contained the relics of an early Egyptian civilization. Through the kindness of Professor Petrie, we were shown rings of gold, set with precious stones, taken from mummy pits, that the professor, who is

Lecturer on Egyptology in connection with this seat of learning, is prepared to prove, according to the interpretation of the hieroglyphic language on the tombs and the inscriptions in catacombs, to be six thousand years old; others he showed us four thousand years old, of much ruder workmanship, proving that the civilization of six thousand years ago was infinitely greater than that two thousand years later. The testimony of this distinguished Oriental scholar, and the teachings of monumental piles and ancient tombs upon Egyptian sands, explode forever the theories set forth by materialistic evolutionists that man has been evolved from a lower to a higher status, but endorse the truths of the Pentateuch that man was a created intelligence, only a little lower than the angels.

But some will say that it is only six thousand years since the creation of man, and many have the impression that the Bible teaches this, and that the theory set forth from the teachings of the hitherto mystic characters to be found in the land of the Pharaohs, is antagonistic to

inspiration. This, however, is not true, for nowhere in the received text is any date fixed as
being the time of man's creation. The first
verse of the opening chapter of Genesis teaches
that in the beginning God created the heavens
and the earth, but in infinite wisdom does not
say when that beginning was; so with regard
to the crowning work of creation, if the re-
searches of modern science revealed the truth,
that it was not only eight or ten thousand years
since the creation of man, but one hundred thousand years, it would still be in absolute harmony
with the declaration of Scripture, the writer of
which was endowed with too much wisdom, by
inspiration, to fix dates. Thus we find on in-
vestigation that the Bible has nothing to fear
from any discovery that may be made along
scientific lines. For nowhere on the oldest
MSS. traced in the original characters on papy-
rus, parchment, or vellum, can any chronolo-
gical data be found fixing the time of man's
creation ?

In one of the departments visited, we were so
fortunate as to see a collection of sculptures that

UNIVERSITY OF LONDON.

would do no discredit to South Kensington; we
do not mean with regard to numbers or variety,
but with regard to excellence of design. One
thing especially noticeable, and rather shocking
to the finer sensibilities of Canadians, was the
nude art on exhibition at the latter place. On
making inquiry of some of our English friends
regarding the propriety of exhibiting nude
sculpture of human figures, we were informed
that the idea had come from France and Italy,
and that in all countries where art had attained
to any degree of perfection, figures were not
draped, and there are some good people who
believe that leaving figures undraped in art
galleries or halls of sculpture, and familiarizing
the masses therewith, robs the more vulgarly
disposed of a curiosity that often leads to gross
crime; and we must confess that along the lines
of social purity the thought is a capital one.

We were also shown seals of great antiquity
and of cunning workmanship, all proving still
further the intelligence of man from the very
beginning, and still teaching, with no uncertain
sound, neither with imperfect knowledge, the

important truth that man has not been evolved from the protoplasmic droplets, that once floated in the primeval seas, to his present degree of intelligence.

Although this is true of man, yet all thinkers along independent lines believe in some form of evolution; and with regard to this planet upon which we dwell, it has been evolved from semi-liquid to consolidated matter, from chaos to order, passing through evolutionary stages and processes, the geological periods proving this: the carboniferous, in which the great coal beds were formed; the ice age, in the breaking up of which the jagged rocks, and many of the rugged hills, were ground down and their pulverized particles strewn over the surface of the earth, preparing it for the reception and growth of the coarser grains. Thus we see the evolving process at work—the consolidation of matter, the formation of inexhaustible deposits of fuel, the preparations for the production of food, all tending from the lower to the higher. But behind this evolution in matter is a supreme something, and this Christian men call God.

WESTMINSTER ABBEY AND ST. MARGARET'S CHURCH.

CHAPTER XIV.

WESTMINSTER ABBEY—ST. JAMES'—CITY TEMPLE.

WE have now come to our last Sunday in London, with its visit to Westminster Abbey, the coronation church of the sovereigns of England, from the time of Harold. Of course, our visit to this grand old historic church, on this day, was to divine service, as the Rev. Canon Farrar was to officiate. The place was crowded almost to suffocation, great numbers thronging the aisles, and compelled to remain standing throughout the entire service. We were surprised at the force and vigour of the distinguished preacher, who is no longer a young man, but who has all the fire of youthful ardour still undiminished. The sermon was so scholarly and eloquent and well arranged, that it almost defied criticism; the hush that fell upon the assembled thousands during its delivery, was so great that one could almost hear a pin fall to the floor.

Between the classical music from surpliced choir and the splendid discourse from one of the most noted ministers on earth, surrounded by the tombs of England's honoured dead, we felt the service to be peculiarly impressive.

Perhaps a few comments on the Abbey, and its history, and what it contains, will be of interest at this stage.

One reason assigned for the conspicuous character of this ancient place of worship is the fact that it is closely connected with the seat of English government. It occupies the site of a chapel, built by Siebert in honour of St. Peter. A church of greater elegance was, however, erected by Edward, about 980, this church being partly destroyed by the Danes. Edward the Confessor founded within the precincts of his palace an abbey and church in the Norman style of architecture, completed in 1065, portions of which remain to this day in connection with this famous place of historic interest. The church, however, was rebuilt by Henry III. in 1220, and finally received the finishing touches during the reign of Edward I.

THE CHAPEL OF ST. EDWARD THE CONFESSOR.

THE CHAPEL OF ... E. & S. IN THE COLLEGE...

The length of the Abbey, including the chapel of Henry VII., is five hundred and thirty-one feet, the width of the transepts two hundred and three feet, and the height of the church one hundred and two feet, and of the towers two hundred and twenty-five feet.

The choir, where all the coronations of English sovereigns take place, was beautifully decorated in the fourteenth century, and contains the tomb of Siebert, King of the East Saxons, Anne of Cleves, and Edmund Crouchback, Earl of Leicester. The northern transept contains the monuments of famous warriors and statesmen, and in the southern transept is the Poets' Corner, containing memorials of most of the great English authors, from Chaucer to Thackeray, and Dickens.

The nave is the most beautiful portion of the building; the monuments in its north and south aisles commemorate musicians, scientists, travellers, patriots and adventurers. The chapel monuments of St. Benedict, St. Edmund, St. Nicholas, St. Paul, St. Erasmus, St. John the Baptist, and Abbot Hislop are all to ecclesiastics and members of the nobility.

Henry VII.'s chapel, with its fretted vault and delicate fan tracery, contains the tombs of many English sovereigns and their children, and also of other individuals of historic fame. In the chapel of Edward the Confessor is his shrine in pure marble, the altar tomb of Edward I., the coronation chairs of the English sovereigns, and the stone of Scone, the old coronation seat of the Scottish kings, upon which some say Jacob rested his head at Bethel.

But the sermon of the Venerable Archdeacon of Westminster, on this our last Sunday in the capital, and the solemn service in connection therewith, amidst the tombs of the illustrious dead, has led the writer into a pardonable description of one of England's most famous landmarks, that has, in part at least, defied the warring elements for nearly a thousand years.

We must now speak of what we heard and witnessed in other churches and places of worship. The pleasure and profit was accorded us of hearing the Rev. Mark Guy Pearse in St. James' Hall, and although a morning service, the place was thronged with eager worshippers.

POETS' CORNER.

Fig. 10.

We do not know the number of persons the vast hall will seat, but it is undoubtedly safe to place it at thousands. We thought at least five, but, of course, could not any more than approximately tell. Be this as it may, however, a vast sea of faces greeted this man, whose name is a household word on both sides of the Atlantic in Methodist circles. His very presence is a benediction, and his countenance venerable and benign. When the text was announced, we noticed nothing striking in the style of presenting the Gospel, and as he advanced both language and thought were surprisingly simple. We looked around the vast audience to see if they appeared interested in the plain discourse, and to our surprise they hung upon the preacher's words. But the writer must confess that he felt puzzled to know wherein lay the man's greatness, but as he waded through the unvarnished gospel, illustrating it by the trees, stones, birds of the air, etc., we at last concluded that his greatness lay in his marvellous simplicity, and when the sermon was completed, we were forced to admit that altogether its beautiful gospel teaching was well calculated to lead men to the higher life.

In the evening of this red-letter Sabbath, we went to the City Temple to hear the celebrated Dr. Parker, author of "The People's Bible." The quaint old temple was filled to the doors, and the Doctor's subject was, "The Origin of Man." We shall never forget his opening words, which were, "Man came from somewhere, that is certain." He then launched out into a very thoughtful and profound discourse, and although the style was not so chaste and scholarly as that of Farrar, yet the profound analysis of his subject proved that he possessed great mental acumen. A marble tablet has been placed in the wall within the ante-room of the Temple, commemorative of the Doctor's faithful pastoral labours during the past twenty-five years. The Temple is not lacking internally in decorative art. It contains costly memorial windows of John and Charles Wesley, Oliver Cromwell, Richard Baxter, John Howe and George White-field. The audiences are very large, showing the popularity of Dr. Parker as a preacher. Indeed, we were surprised at the size of the congregations in all the cathedrals and churches

INTERIOR OF EXETER HALL.

which we visited while in the Old Land. This
is surely evidence that Christianity is laying
hold of the masses in thoughtful England,
and the outcome of the matter will be the
hastening of the day, foretold by prophet and
seer of old, in the which the Son of God shall
have the heathen for His inheritance, and the
uttermost part of the earth for His possessions,
for England is not slow to make use of her con-
secrated and gifted sons in heathen lands, to
help bring the pagan masses from out the
shades of overspreading night.

CHAPTER XV.

HORSE-GUARDS AND ADMIRALTY.

BEFORE leaving London, we must make mention of a flying visit to the Admiralty and Horse Guards. The buildings of the former were erected in 1726, by Ripley, forming three sides of a quadrangle, with a screen and gallery towards the street, designed by the Adams Brothers, Architects of the Adelphi Theatre and Portland Place. Capping the screen are sculptures of sea-horses. The official residence of the First Lord of the Admiralty is here. In the residence of the Secretary are portraits of persons who have filled that position, from the earliest down to the latest officials. Before the introduction of the electric telegraph, the semaphore on the roof of the buildings was one of the sights of London. Corresponding semaphores were erected at short distances, on the road to Portsmouth, and intelligence was trans-

mitted by a system of signalling, similar to that
in use on railways.

The Admiralty, properly speaking, is the
great electric button, the touch of which regu-
lates the movements of the enormous naval

NEWGATE.

armament of the Empire in all parts of the
world.

The building of the latter, the name of which
has been taken from the Mounted Household
Troops, always on duty there, is erected on the
site of the old tiltyard of Westminster, so
renowned in the annals of the Tudor times.

There tilted Philip Sydney, the Earl of Leicester, Sir Christopher Hatton and brave old Sir Christopher Lee, who had lived eighty years and served five English monarchs. There Queen Elizabeth sat enthroned in royal state, and Duke d'Anjou, son of Catharine de Medicis, who had crossed to England to ask the maiden queen to marry him, took part in a chivalrous pageant.

In the alcoves that guard the entrance to the building proper, sit two mounted guardsmen, each six feet tall, with steel breastplates and polished helmets, like uniformed marble statues —we doubt not, as valiant warriors as ever tilted in Tudor times in the old courtyard.

The present building was erected in 1758, at a cost of over £30,000 sterling. The clock is a masterpiece of skill, illuminated by reflection, and is one of the best timekeepers in the kingdom. The military guard on duty is provided alternately by the two regiments, the Life Guards and the Horse Guards, the change of which every morning at eleven o'clock is a most interesting sight.

CHAPTER XVI.

WESTMINSTER HALL—TOWER BRIDGE.

IF we could not leave London without giving
some description of the above places of interest,
it would be almost criminal to omit Westminster
Hall, second in interest only to Westminster
Abbey and the Tower. It is now the public
entrance to the Houses of Parliament and the
courts of law. The Hall was first built by William
Rufus, and rebuilt by Richard II. in 1397,
the clerk of the works being no less a personage
than Geoffrey Chaucer, the father of English
poetry. Under his superintendence the height
of the Hall was increased and the splendid
timber roof added, which may be seen to this
day. Later improvements have been made,
adapting it for being the grand vestibule of the
Houses of Parliament.

This is one of the largest halls in the world
with roof unsupported by pillars ; the length is

240 feet, the breadth 70, and the height 92. The timber roof is constructed of chestnut wood; the ornamentations of the projecting beams are angels supporting shields charged with the arms of Richard the Second and Edward the Confessor, and on the stone frieze beneath the windows are sculptures of a hart couchant and other devices.

The roof was repaired and extended in 1820 with oaken beams taken from old ships of war at Portsmouth. The first English Parliament sat here; royal festivities, among them the coronation banquets, were held here, the last being that of George the Fourth, when the champion of England rode into the Hall and threw down the glove against all comers.

Charles the First was tried and condemned, Cromwell was proclaimed Lord Protector, and a few years later his head was exposed here on the point of a pike. Here, too, were tried and condemned William Wallace, Lord Cobham, Sir Thomas More, the Protector Somerset, the Earl of Essex, Sir Thomas Wyatt, the Earl of Strafford, Guy Fawkes, Lord Balmerino and Lovat, and Lord Ferrers, who murdered his steward, and last but not of least importance was here held the famous trial of Warren Hastings. In the Hall are statues of James the First, Charles the First, Charles the Second, William and Mary, George the Fourth and William the Fourth.

On the 16th of October, 1834, the Houses of Parliament were swept away by the fire fiend, and the first stone of a new structure, on a gigantic scale, was laid in April, 1840. The site occupies an area of eight acres, and the buildings are 900 feet in length and 300 feet in width. A terrace 940 feet long and 33 feet wide extends along the river, towards which is the principal façade of the building, decorated with the statues of the kings and queens of England, from William the Conqueror to Queen Victoria, and panelled sculpture representing coats of arms and royal devices. The style of architecture is the finest Gothic, and the stone used in the construction of the buildings is magnesian limestone, from Yorkshire, but an unfortunate blunder has been committed in the selection of this stone, which is already beginning to suffer from the effects of a London climate.

In the interior, Caen stone has been used, and the river terrace is of Aberdeen granite. There are 11 open courts, 1,100 apartments and 100 staircases. The great Victoria Tower at the southwestern angle is 331 feet high to the top of the metal crowns on the angle turrets, 75 feet square, and rises over four pointed arches, 60 feet in height. There is a spiral staircase of 553 steps to the upper floor of the Tower. The flagstaff is of wrought iron, 120 feet high, 2 feet in diameter at the base, and 9 inches at the summit, surmounted by a copper-gilt crown, 3 feet 6 inches in diameter, and 5 feet 6 inches high.

The royal entrance is in the Tower. Over the Central Hall, is a very elegant Gothic spire, rising to the height of 261 feet, and at the northwestern angle of the mighty structure is the clock tower, 40 feet square, 320 feet high, containing a clock that has four faces, each dial being over 22 feet in diameter, all illuminated at night. The hours are struck on a bell, called "Big Ben," weighing nine tons; the quarters are chimed on eight smaller bells. The interior of this magnificent palace is gorgeous in the extreme.

THE MANSION HOUSE.

The House of Lords is a superb hall, 97 feet long, 45 feet wide, and the same in height; it is lighted by twelve painted windows, with portraits of England's monarchs. The throne has a gorgeous gilt canopy, and the decorations of the House are extremely ornate. Between the windows are niches in which are statues of the barons who compelled King John to sign the great charter of English liberty, and on the lofty arches at each end are frescoes, such as the Baptism of Ethelbert, Edward III. Conferring the Order of the Garter on the Black Prince, and Judge Gascoigne committing Prince Henry to the Tower. These frescoes are above the throne, and at the opposite end are symbolic figures, such as Religion, by Horsely, and Justice and Chivalry, by Maclise. Before the throne is a cushioned seat, the famous woolsack, on which the Lord Chancellor sits. The seats for the peers are covered with red Morocco.

At the end of the House, opposite the throne, is the bar, a space to which the members of the House of Commons are admitted to hear the royal speech at the opening of the session of

Parliament. An ornamental gallery is beneath the window, and there are reporters' and strangers' galleries. The ceilings and walls are adorned with heraldic ornaments, symbols, devices and monograms.

In the Peers' Robing-room are two grand frescoes by Herbert, Moses Bringing Down the Law, and the Judgment of Daniel. The Peers' Lobby is divided from the House by massive brazen gates, and beyond it is the Peers' Corridor, in which are eight frescoes, The Funeral of Charles I., Expulsion of Members of one of the Oxford Colleges for Refusing to Sign the Covenant, Defence of Baring House by the Cavaliers, Charles I. Planting his Standard at Nottingham, Speaker Lenthall Defending the Rights of the House of Commons against Charles I. when he attempted to Arrest Five Members, Departure of London Citizens to Assist the Garrison of Gloucester, Departure of the *Mayflower* for New England, Parting of Lady Russell from her Husband, Lord William Russell, before his Execution.

The Octagonal Hall is 60 feet in diameter

and 75 feet high, with a vaulted stone roof, ornamented with Venetian Mosaic representing the heraldic symbols of the Arms of England. Above the doorway are pictures in Mosaic. In a niche at the side of the door is a statue of Earl Russell, by Bohme. In the Waiting Hall are frescoes representing scenes taken from the works of British poets, as follows: Patient Griseldo, from Chaucer; Red Cross Knight Overcomes the Dragon, from Spenser; Lear Disinheriting Cordelia, from Shakespeare; Satan Touched by the Lance of Ithuriel, from Milton; St. Cecilia, from Dryden; The Thames, from Pope; The Death of Marmion, from Scott; The Death of Lara, from Byron.

In the Commons Corridor are also the following frescoes: Alice Lisle Concealing Fugitives after the Battle of Sedgemoor, The last Sleep of Argyle, Acquittal of the Seven Bishops, The Lords and Commons Offering the Crown to William and Mary, General Monck Announcing his Support of the Liberty of Parliament, Disembarkation of Charles II., Execution of Montrose, Jane Lane Assisting the Flight of Charles II.

The House of Commons is less richly ornamented than that of the Peers. It is sixty-two
feet long by forty-five in width. The roof is of
ground glass, above which are the gas jets used
for lighting the place, an arrangement by which
the atmosphere is kept pure. There are twelve
windows, on which are painted the arms of
boroughs returning representatives to the
House. The Speaker's chair is at the northern
end, and in front of it is the table on which the
mace is laid while the House is in session.

The royal entrance in the Victoria Tower is
by a Norman porch with vaulted roof, supported
by a beautiful central pillar; leading from this
is the royal staircase, fifteen feet wide, with
twenty-six steps of Aberdeen granite, and
lighted by painted windows, with portraits of
Edward the Confessor and Queen Victoria. This
staircase leads to the Queen's robing-room, richly
decorated with frescoes by Dyce, representing
the Virtues of Chivalry, as illustrated in the
Arthurian legends. From this room the Victoria
gallery, one hundred and ten feet long and
forty-five in height, leads to the House of Lords.

NATIONAL GALLERY.

The floor is formed of beautiful Mosaics, and the ceiling is wainscotted and richly gilded. On the walls are two large paintings, each twelve feet by forty-five, one representing the Death of Nelson at the Battle of Trafalgar, the other the Meeting of Wellington and Blucher after the Battle of Waterloo.

In the Prince's chamber there is a marble group, by Gibson, of Queen Victoria on the Throne, supported by Mercy and Justice. Three painted windows show the Rose, Thistle and Shamrock.

At the upper end of Westminster Hall is St. Stephen's porch, leading to the hall of the same name, with a fine gallery, ninety-five feet long and thirty feet wide, in which are statues of John Hampden, Seldon, Sir Robert Walpole, Lord Chatham, William Pitt, Henry Grattan, Lord Clarendon, Lord Falkland, Lord Somers, Lord Mansfield, Fox and Burke. A door, near the southern end of the palatial pile, gives access by a flight of stone steps to St. Stephen's crypt, a remnant of the old building, and now used as the chapel for the residents of Westminster

Hall. The chapel is ninety feet long, twenty-eight feet wide and twenty high. It is lighted by seven stained glass windows, representing the ministry and martyrdom of St. Stephen, St. John, St. Catherine and St. Lawrence, and the legend of St. George and the Dragon.

The east wall is panelled with full-length figures, on a gold ground, of St. Peter, St. Oswald, St. Etheldreda, St. Edmund, St. Margaret of Scotland, and St. Edward the Martyr. The columns of the chapel are of polished Purbeck marble, and the place will seat about two hundred and fifty persons.

After leaving this magnificent pile of masonry, in which the laws are framed that regulate the actions of many millions of the human race, and in which England's most distinguished sons, both Lords and Commoners, meet to debate and legislate, we bent our steps in the direction of the new Tower Bridge that had been but recently constructed, at a cost, we were informed, of one million pounds. The structure spanned the river opposite the Tower of London, and was built of iron and steel, with two towers, one at each end

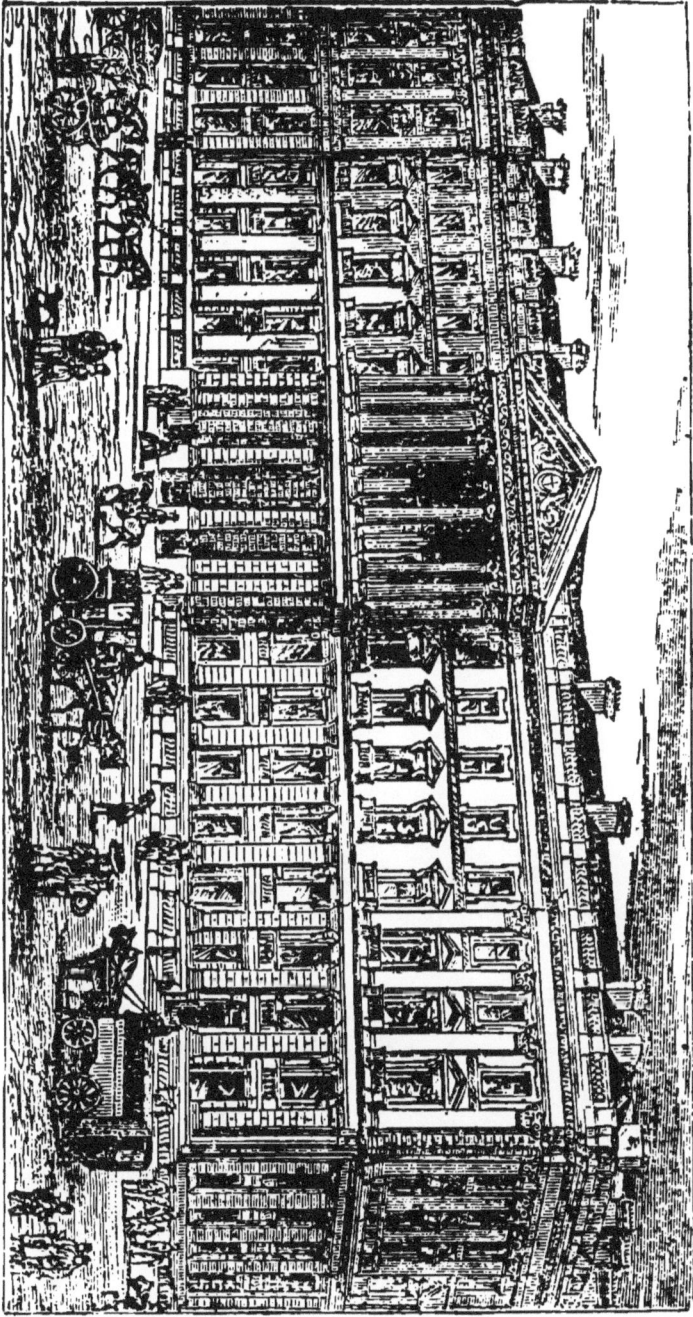

GENERAL POST-OFFICE—NEW BUILDINGS.

nearly one hundred feet in height, so that when vessels were due to pass that point in the river the bridge was raised, by powerful machinery, to the top of the towers, traffic going on all the while, as they were each equipped with ponderous elevators that lifted both foot-passengers and vehicles up until on a level with the bridge proper. These elevators kept continually going while the shipping with tall masts passed underneath, thus preventing the suspension of traffic for a single moment. The structure was opened to the public by His Royal Highness the Prince of Wales, in the presence of a distinguished company, and with imposing ceremony.

We half suspect that there existed a double purpose in the construction of this massive fabric: the first, to facilitate traffic and accommodate the public; the second, to act as a boom across the river, for the bridge could be lowered to the level of the water, if need for this course existed, so that if in war-time hostile ships should force the batteries at the mouth of the river, they could not possibly ascend it farther into the city than the Tower Bridge, where batteries erected on the famous tower hill, could play with deadly effect upon them, and the structure of the bridge is such, and its strength so great, that the fire of ordinary guns would have but little effect upon it.

CHAPTER XVII.

BUNHILL FIELDS.

THE last place visited in London was Bunhill
Fields, the God's-acre of Nonconformists, situated
near the City Road, where upon the tombstones
are recorded some of the most illustrious names
in Nonconformist annals.   Dr. Thomas Goodwin,
who attended Cromwell in his dying hour, lies
here ; also George Fox, the founder of the
Quakers ; General Fleetwood, Oliver Cromwell's
son, and John Owen, the independent divine.
There are also three names that, no matter
what may be our theological convictions, we
cannot pass over in silence, namely, John
Bunyan, author of " The Pilgrim's Progress,"
Daniel Defoe and Isaac Watts.   Adjoining the
cemetery is the barracks of the London militia,
and next to that the artillery ground, where the
members of the Honourable Artillery Company,
the oldest of the volunteer corps of the city,

11

first incorporated in 1585, and known as the City Trained Band, have their drill-ground and armoury. Opposite the military barracks is the Wesleyan Chapel, the first stone of which was laid in 1777 by the saintly and immortal John Wesley, who preached there during the later years of his life, and was buried there. In front of the building which stands back from the road is a monument to the memory of his noble mother, Susannah Wesley. In the chapel is a tablet in loving memory of Charles Wesley, the sacred bard of Methodism. Leaving those scenes, sacred by a thousand hallowed associations to every Methodist heart, and nearly at the close of another memorable week, we returned to our lodgings to make preparation for leaving the vast city in order to visit other places of interest in the country.

CHAPTER XVIII.

WATERLOO STATION—SUNNINGDALE PARK—VIRGINIA WATERS—WINDSOR PARK.

IN harmony with our plan, next morning we crossed the Thames by the Waterloo Bridge, one of the finest that spans the historic river, and reached the station of the same name, about half an hour before the train started, giving us an opportunity of looking around for a time. Passengers are not allowed to enter any of the cars until a very short time before the train leaves the station, the approaches to the different trains standing in the depot being guarded by a strong fence with iron gates, at each of which is stationed a guard. Someone said in our hearing that a royal carriage was attached to the train; but in looking over all the carriages connected therewith we noticed very little difference in their construction, and concluded it was a mistake, so when the guardsmen gave permission we took a seat in one of the cars, in

Canadian parlance, and very soon were career-
ing over a lovely country towards Sunningdale
Park, one of the most beautiful places in
England, and, best of all, the rumour about

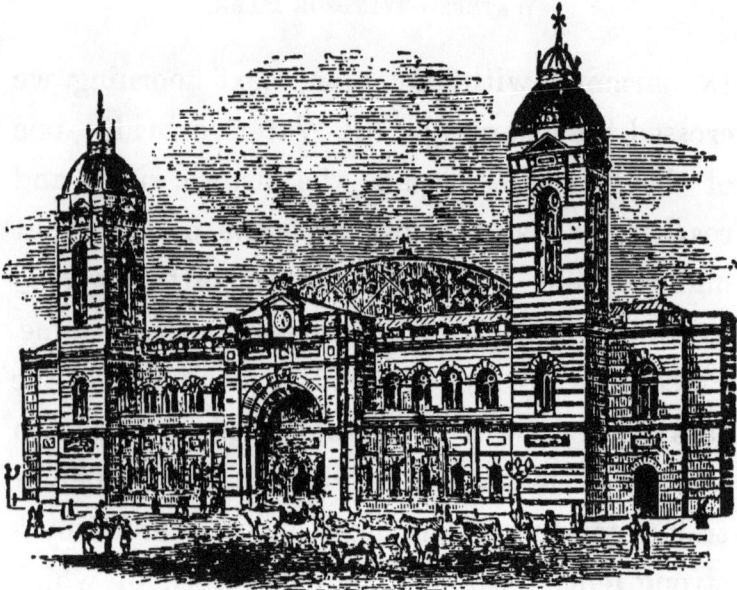

AGRICULTURAL HALL.

the royal carriage was destined to prove an
actuality, for as the train pulled up at the
station the Prince of Wales, heir-apparent to
Britain's throne, accompanied by the Duke of
Cambridge, stepped out on the platform. We

had the misfortune to alight on the wrong side of the train to see very much of the distinguished personages, but caught a glimpse of them for one brief moment. His Royal Highness is a pleasant-looking man, rather stoutly built and of benign expression; whilst the Duke is every inch a soldier. As he stood for a moment it was " at attention," and as he turned toward the carriage in waiting it was " to the right wheel."

We felt especially delighted to have had the honour of travelling thirty miles on the same train and then seeing our future sovereign, if God sees fit to spare him. If our readers will permit the use of familiar and popular language, we may say that the Prince is " a jolly good fellow." Some beautiful stories are told of his geniality and kindness of heart, and intelligent Englishmen informed us that he was the most popular man in the country, and that if Britain were created a republic to-morrow, the first President, by acclamation of the people, would be His Royal Highness Albert Edward, Prince of Wales. England, however, does not desire a

republican system of government, having learned
long ere this that the safest form of government
on earth is her own limited monarchy.

Republicans speak of the fact that their President is so off-hand and approachable, but judging from the appearance of the Prince of Wales,
and also from the testimony of his countrymen,
except on state occasions, where court etiquette
demands a certain dignified reserve, the Prince
is as approachable, and as genial and off-hand as
the chief magistrate of any republic on earth.

Mr. Frederick Thorn was waiting at the
station to welcome the writer. We had formed
the acquaintance of the brother-in-law of this
gentleman in Canada, who had written Mr. Thorn
of our arrival in the Old Land, and of our lodging
place in London; the latter, on learning this,
called on us at our boarding place in the capital,
and invited the writer to visit his home, at the
place before mentioned. It was peculiarly pleasing to do so, not only because of the kind invitation, but also because of the fact that in this
beautiful suburban place many of the nobility of
England resided, and we were anxious to acquire

more information regarding their mode of life, morals, influence, etc., and Mr. Thorn, who was much respected in the locality for his integrity and Christian manliness, as well as the fact that he was a member of the Royal Horticultural Society of England, and an intelligent, genuine Englishman, was just the person to place us upon the proper track for the acquisition of the desired information. Through his kindness, we were shown through the beautiful mansion of Major Joicey, with its pillared front, even to the vaults in the cellar, where the gold and silver plate of the family is preserved.

The Major is reputed to be enormously wealthy, and his estate is indeed a veritable paradise— especially the spacious grounds surrounding the family seat, with their drives, flowers, miniature lake, lordly trees, their branches in many cases supported by massive iron chains to prevent their splitting off the parent stem. The neat cottages for the workmen, the wooded parks where game abounds, the carefully trimmed hedges, and many other things evidence cultivated taste and pardonable ambition.

The Major is highly spoken of in the community, and we were informed that in all probability he would soon be made a Peer of the Realm.   Some of his neighbours were Lords, and we may add just here that we went to England

BUCKINGHAM PALACE—GARDEN FRONT.

prejudiced against the peerage, and the House of Lords, and after learning more of their high social position, that is in itself a preventative of low actions, from the fact that family pride will not allow them to stoop to the level of the criminal, for they are of noble birth, and this very birth is, to a certain extent, a guarantee of good behaviour; but also many of them—we repeat, many of the peers of England are men of integrity, men of meritorious manliness, men of marked mental ability, as seen in such examples as Lord Salisbury, Lord Rosebery and others; and women like Lady Henry Somerset; so that we left the Old Land with quite different feelings to those we possessed on entering it, and now emphatically assert the truth that from this time forward we have entertained the kindliest feelings towards the Peers of the Realm, and also towards the House of Lords.

As a rule, the members of the British peerage have fine homes, with every equipment that cultivated taste can conceive of. This can also be said of nearly all wealthy Englishmen, who in most cases belong to the better class of society.

On the estate of Major Joicey they had their own fire brigade, composed of the hands working thereon, with fire-engine, hose-reel, hook and ladder wagon, all evidently of the most costly pattern, with uniforms for not only active service at fires, but also elaborate uniforms for parade occasions. When it is understood that there are scores of employees on an estate of this kind, the thoughtful reader can at once conceive how easy it is to draft a sufficient number of the young men, for the formation of such a brigade and in addition to the equipments mentioned above, at every important point was constructed a fire alarm, of the most approved pattern, so that a conflagration at the most distant part of the estate could be made known in a moment at the head office, and a brigade that would do no discredit to the metropolis itself, borne by the best horses in the kingdom, would soon be on the ground. The Major's son was chief of the department. All these costly appliances and equipments give some idea of the enormous wealth of the owner of such a property.

After describing at some length things of interest on this typical estate of an English gentleman, we must not forget the gardens, in which are the conservatories, wherein abound flowers and fruits that grow in all parts of the world, even to the tropics. In the glass structures that contain tropical plants and fruits, are thermometers, so that the heat from enormous furnaces can be regulated, to bring the temperature to the tropical point. Coming, as we did, from a cool northern climate, we could not stand the insufferable heat of those conservatories for more than a minute or two. In the fruit departments were luscious ripe peaches, plums, apples, strawberries, raspberries, and almost everything that could tempt the appetite of the most fastidious. Through the kindness of our host, Mr. Thorn, we were invited to eat all the fruit we possibly could, and upon the peaches especially we lavished a peculiar degree of affection. Our friend, whose name we have just mentioned, had charge of this department with a large and competent staff of assistants under him.

At the Orchid Exhibition in London, some time ago, we understand that Mr. Thorn took the silver medal of the Royal Society.

The arrival of another Sabbath put a stop to sight-seeing for the time. We spent a quiet day, attending service but once, at the Wesleyan chapel, about a mile away. The sermon was delivered by a good old English local preacher, and with considerable power and intense earnestness, and we must acknowledge that although we had listened the two previous Sundays to some of England's most famous preachers at the capital, yet the spiritual profit was not greater than in listening to the burning words of this veteran layman, who, without any earthly reward, had for more than twoscore years held up the banner of God. Methodism in England owes much to these earnest, self-sacrificing and devoted men, as it also does in Canada and elsewhere.

The Sabbath, with its opportunities for quiet meditation having passed away, next morning we availed ourself of the privilege of visiting Virginia Waters, a delightful resort for the better class of society. On the shore of this lovely stretch of water, Her Majesty the Queen has a boat-house and suite of pleasure apartments, where members of the Royal family while away many a pleasant hour.

Up the lake some considerable distance is an extensive ruin, consisting of pillars, broken arches

and fragments of crumbling masonry that once
had taken the form of walls, and at the water's
edge were stone steps with end caps of solid
blocks of chiselled granite. On the face of one
was a lion's head, but the destroying hand of
time had crumbled the device on the other. We
made strenuous efforts to ascertain the age of the
interesting ruins, but nobody seemed to know.
One said he thought the stones had been brought
from an ancient ruin in Greece by one of the
Georges, and built here, but this seems highly
improbable, and on closely examining the standing pillars, as well as the masonry on the water's
edge, we found the design very similar to that
of the Roman ruins in other parts of the country,
and consequently arrived at the conclusion that
this was an ancient Roman ruin, perhaps sixteen
hundred years old.

The next day after visiting Virginia Waters
and the ancient ruin, which is the wonder of
archæologists, we visited Windsor Castle, the residence of Her Most Gracious Majesty Queen Victoria. The walk through Windsor Park toward
the Castle was delightful; what with its quaint

old lodges and majestic trees, it is something one does not see every day. Game also abounds; so numerous were the rabbits that no man could number them, and deer, thousands upon thousands. Those who think all the deer are in the northern forests of our vast Dominion, would do well to walk the seven miles from Sunninghill Lodge gate, through the Park to the suburbs of Windsor town, and we venture to say that the result of the ramble would be that their minds would be completely disabused of such erroneous notions.

We had not proceeded more than a couple of miles when we met Prince Christian, the Queen's son-in-law, and his daughter, who were out driving. The Prince is an elderly gentleman, and rather venerable-looking, and is, we were informed, a man of exceptional character; we would judge from his appearance that this is substantially correct.

Another mile brought us to the statue of the late Prince Consort, around which is a drive, and a beautiful thing was told us of our beloved and widowed Sovereign, that every time she is driven in the Park, she requests her coachman to drive around the statue of this virtuous Prince, who during his lifetime was so greatly beloved, not only by a faithful wife and devoted children as a husband and father, but also by a nation, as a man of sterling integrity and exalted character. Her Majesty's undying love for her departed husband is one of the evidences that true love is an eternal principle.

Another mile or two brought us to the copper horse of King George, or more properly speaking, an equestrian statue of that monarch, in solid copper. The figure of the horse stands upon a pedestal of stone, and is of a considerable height and quite imposing; the sculptor, however, in finishing his work, forgot to add the stirrups, and when the statue was erected and the public noticed the omission, the newspapers at once took the matter up, and so severely criticised it because of this defect, that the designer committed suicide, so great was his chagrin.

The above statue is at the end of the avenue, straight as an arrow, which is called the long walk; it is about three miles in length from the copper horse to the gates of Windsor Castle, and is bordered on either side by double rows of lordly elms three hundred years old. The avenue itself is fully one hundred feet or more in width, and is, we believe, the most beautiful this side the streets of gold, spoken of by the seer of Patmos.

CHAPTER XIX.

Windsor Castle—Eton College—Ascot, etc.

ON passing into the courts of the Castle we found armed and helmeted Grenadier guardsmen parading everywhere—fine, strapping fellows, none of them less than five feet ten inches in height. With their great shaggy busbies they present a very fine appearance. The standard was formerly six feet; but as men are growing smaller physically with increasing wisdom, the military authorities could not secure a sufficient number of six-footers to keep the regiment at full strength, and accordingly were compelled to lower the standard to the first-mentioned figures.

The writer had a conversation with one or two of the men who were on duty, and was privileged to see them change guard, which they did with marvellous precision and exactitude. One of the men with whom we conversed showed us his rifle, which we examined closely. It was of

12

the Lee-Metford magazine pattern, with short
sword-bayonet, and is doubtless one of the best
weapons ever manufactured:

The guardsmen were quite communicative, and
apparently as intelligent as they are fine-looking. The members of this regiment are doubtless among the finest troops on earth. The corps
has its barracks in connection with the Castle,
parts of which are very old, bearing the date of
1583.

The Royal Chapel is about three hundred years
old. The architecture of the older portions is
in the Elizabethan style. The tower of the
Castle is not only lofty, but very strongly built,
being circular in form and mounted all around
with cannon, and apparently constructed for
withstanding a close siege. Heavy guns are
mounted on the parapet on the side of
Eton College. The Castle is indeed a splendid
fortress.

At the time of our visit the Queen was absent
on the continent, and consequently her private
apartments were not open to the public; but in
those apartments, we learned, what, of course,

we would expect, that the furnishings and drap-
eries were exquisite; but the plate and most
valuable appurtenances, which are kept in the
vault, cannot be seen, we understand, on any
occasion, and the Crown jewels, which we described in an earlier stage, are preserved in the
Tower of London.

We had an excellent view of old Eton, with
its famed college, from the courtyard of Windsor,
and also had the pleasure of meeting some of the
students, who, by the way, are great favourites
of Her Majesty. The reason assigned for this
is, that when an attempt was made some years
ago, by a crank, on her life, the weapon in the
hand of the would-be assassin was struck aside,
when it was about to be discharged, by one of
the Eton boys, who had come on the scene at
the opportune moment, and ever since the boys
have been favourites at the Castle. It was told
us that there were nine hundred attending the
College, and all that we met, no matter how
small they were, wore tall silk hats. All the
students, by a stipulation in the rules of the
College, are expected to conform to this practice.

Eton College is a time-honoured institution, and is doing a great work among the more respectable classes of society in educating their youth to be intelligent and useful citizens.

After leaving the Castle we took a walk through the city, or at least a portion thereof. It is doubtless quite old, as the buildings and streets evidence, and a considerable amount of business seems to be transacted therein.

The country around Windsor is as beautiful as a park, trees and buildings and green fields beautifully alternating, making every prospect to please the senses. We do not wonder that Englishmen love England, for it is indeed a fascinating country. Of course, we must say, in adherence to truth, that there are tracts almost worthless, and we passed over thousands of acres of such land stretching away off in the direction of Aldershot Military Camp; but this is the exception, not the rule.

After leaving the Royal city of Windsor, and resting for a short time, for we found revelling amid historic scenes and sight-seeing very exhaustive, especially so to us in our invalid

condition, we visited Ascot, chiefly important
for being the scene of the great national races.
Indeed, one week after we left for Birkenhead
the races came off. The Prince of Wales and
many of the nobility were to honour Ascot
with their presence, and even Lord Rosebery,

GENERAL POST-OFFICE—OLD BUILDING.

a man of marked ability and excellent parts,
whose only hobby is a fast horse, was to pay
his compliments to this smart little town on the
occasion of the races. We may also state that
an invitation was given the writer to remain
another week to witness this great event, but
as our aspirations did not lie in the direction of

fast horses, we declined as graciously as possible the kind invitation.

The town of Ascot, although not large, possesses some substantial buildings, and we were shown beautiful cottages, owned by workingmen, principally mechanics, who had by the labour of their hands, and the blessing of divine providence, saved enough money to purchase a plot of land and erect thereon those lovely little houses, showing that honest, industrious workingmen can make their way in England as well as elsewhere; indeed, while in conversation with a police officer in London, he said that in his opinion they were more democratic in England than we in America. But if we cannot exactly agree with our friend the police officer, we can at least respect his opinions, and one thing is morally certain, that within recent years the English people have made giant strides in the direction of true democracy—not the socialistic levelling up which fanatics and communists teach, but that condition of things which delegates to the labourer power to purchase and to hold property on the condition of his paying four dollars and eighty-six cents on the pound.

Before leaving Sunningdale Park, we spent a day or two rambling in the woods and English country lanes, the former all resembling parks, and the latter!—no language is rich enough in terms to describe their glorious, romantic beauty, with hedges on either side in full bloom, and the lane shaded by the overspreading boughs of enormous trees. We shall never forget, while memory keeps her throne, our visit and walks amid the gorgeous scenery and delightful landscapes of southern England.

And now, the days of our sojourn here having ended, we bade adieu to the Thorn family, who had been more than kind to us, and taking the morning train arrived long before noon at the metropolis; and proceeding at once to our old lodging place, packed up our luggage, took a short rest, had dinner, and then boarded a train at Euston Station, and after an uneventful journey, consisting of passing through a long dark tunnel just outside the city, and then through gardening districts and finally through the productive agricultural sections farther north, in something over four hours we arrived at Lime Street Station, Liverpool, and at once proceeded to the Salisbury Hotel, where we took a nice room, and after a day or so felt quite at home.

When the building, which is most substantial, was being erected, the owner, Mr. Lloyd, waited upon Lord Salisbury, now Premier of England, and asked him for the privilege of using his name in connection with the hotel. The Peer replied that if he would not dishonour it he would be most happy to grant his request. Mr. Lloyd promised, and consequently the place was named the Salisbury House.

After spending a short time in this comfortable temperance hotel, we learned that the owner was in the habit of coming at six o'clock every evening to one of the rooms to hold a short religious service, where all the servants in connection with the establishment were expected to gather, and also any of the guests who desired to do so ; and accordingly the writer, glad of an opportunity of attending divine worship amid the dissipations of travel and hotel life, proceeded to the appointed room, and after an impressive little service, which was much enjoyed, made the acquaintance of Mr. Lloyd, who very courteously invited a Presbyterian missionary just returning from India with his family, who also, like the writer, was suffering from nervous prostration, together with the author, to see the sights of the great city with half a million of inhabitants.

CHAPTER XX.

LIVERPOOL—BATTLEFIELD OF EDGE-HILL.

WE were delighted with the appearance of the public buildings in Liverpool, which we began to realize was like Tarsus, no mean city. Among the many places visited was the Young Men's Christian Association building, near the central part of the city, a magnificent and stately structure. In the gymnasium, in connection therewith, we noticed a large number of young men, finely built and of splendid physique, running around in a large circle on an easy trot. We asked our guide how it was that they were such strapping fellows. He answered, pointing to the runners, because they do that. It was only recently that we recalled these words, and now every day take a run in the cellar, with the result of a marked improvement in health. It is said that a man is a fool who will not put in practice what he learns at infinite cost. We learned this, together with a few other things, at considerable cost, and feel determined in the future to utilize the knowledge acquired in the past.

During our ramble around the city, we learned a good deal of India from our Presbyterian friend, and among other things the fact that his last congregation in that far-away land was Her Majesty's 78th Royal Highlanders.

When our round of sight-seeing was completed, Mr. Lloyd invited us to his home, and introduced us to his family, where after spending a very enjoyable half-hour, we were asked to partake of luncheon. When that pleasant duty was performed, we took our departure for the hotel, feeling that we had fallen in with a very courteous and hospitable gentleman.

One of the places visited the following day was the gallery, containing what is known as the Walker collection of Art, where many noble pictures were exhibited, among them one monster oil-painting of the Battle of the Nile, and so intensely realistic as to prove the excellence of the work.

We were shown the home of the Rev. Charles
Garrett, that distinguished and well-known
English divine, and the room in which he was
even then thought to be writing, and also the
chapel in which he often preached.

An old church-yard within the precincts of
the town also proved to be a place of great
interest; many of the tombstones had fallen
and were lying upon the ground. On some
were engraven names that are recorded in history, and altogether the apparent age of the
place made it quite interesting.

Hearing that the famous battlefield of Edgehill was not far distant, we determined to visit
it, and found that the city had extended to its
limits, and that it was all laid out in streets
and mostly built over; but having carefully
studied the history of the battle, it was interesting, and yet sad, to walk over the ground and
think that long ago the remains of many a
Parliamentarian, and many a Royalist, had
crumbled into dust on this very spot. It was
here that Lord Essex and the King measured
swords in the opening of the terrible Civil War.

It was on this soil, baptized by the life's blood of some of England's most valiant men, that Sir Faithful Fortesque (inappropriately named, should have been "Sir Faithless"), deserted his commander with his entire regiment, throwing the Parliamentary army into confusion, whilst the Royalist horse, with irresistible valour, drove their opponents headlong from the field. But in return for this, the reserve of Lord Essex completely shattered the Royalist infantry, and had it not been for the opportune return of Prince Rupert from the pursuit of the broken regiments, the Royalist cause would forever have been shattered to atoms; as it was, however, the shades of evening gathered upon an undecisive battle with the advantage but slightly on the side of the King.

CHAPTER XXI.

Birkenhead —Dock-yards—British Navy.

The next expedition after visiting the battlefield of Edgehill, which was planned with our
friend the Presbyterian missionary, was across
the Mersey River to the city of Birkenhead, to
visit the great dock,yards where many of Her
Majesty's warships were built. He, however,
did not think that we would be able to gain
admission, but thought the expedition would,
at all events, do us good. Accordingly, we had
a lovely trip on the river ferry to the great and
important city on the farther shore. After
landing, we proceeded at once to one of the
entrances to the extensive ship-yards, where
we sought admission, our friend keeping back
a little, as he was an American citizen, really,
though so many years in missionary work in
one of Her Majesty's possessions. The gate-
keeper asked the writer if he were a British sub-

iect, and the answer was, " With all our heart ; "
but he informed us that we had better proceed
to the head office. This was not very encourag-
ing, and our American friend said, " You will
find what I told you is correct, we will not get
admission." We, however, next tried the main
entrance, and was there informed that it was
only by going to the office upstairs and pro-
curing a passport, that we would be admitted.
Our friend said, "You can go if you wish, I have
given it up." We went, however, and asked the
official in charge if we could go through the
yards, and if he would grant us passports. He
put a few questions, such as the following :
" Are you a British subject ? " " Is your friend
the same, or is he friendly to our nation ? "
" What use do you purpose making of the information a quired ? " etc. After answering the
above questions as best we could, and doubtless
impressing the official favourably with our good
looks, he cheerfully and readily granted us the
coveted passports, and also a boy to act as
guide. Going down stairs the news of our suc-
cess was a pleasant surprise to the worthy

missionary, who had declared that he was sure we would fail.

It is scarcely necessary to say that the couple of hours spent within those walls were most interesting and profitable. Enormous stacks of steel and iron, piled here and there through the yards, gave some faint conception of the quantity of material used in the construction of vessels of different patterns.

Several torpedo boats were in the docks for repairs, and having never before seen vessels of this kind, we were peculiarly interested in inspecting them. They are not large, as this would be a disadvantage rather than otherwise, impeding rapid motion, upon which everything depends. From the shape of all these boats, and from the fact that they are furnished with twin screws, speed seems to be the main object, as they are capable of being propelled at the rate of from twenty to thirty miles an hour. In the bow is a dynamite gun, or more properly, perhaps, a tube from which the torpedo is discharged. The boats are all armoured, and immediately over the tube port is a heavy steel gate or door which is closed the moment the projectile is discharged, to prevent a shot from dismounting the gun. Every part of the machinery is protected, and nothing on board one of these boats can be injured by exploding shells.

After examining one of the torpedo boats very closely, for they are all similarly constructed, we were shown the foundation of an enormous battle ship which had just been laid, and which was designed, we were informed, to be one of the largest in the British navy. It was being built very much on the principle of the modern fort.

A few thoughts just here will be in order in connection with the navy of England. There are, at the present time, three hundred ships of war of all classes in commission; one hundred more in course of construction at the various dock-yards, that could be placed under commission in a month or two; another hundred—some of them non-commissioned at present, some of them ordered to be constructed at as early a date as possible—all of which could be com-

13

pleted in a few months. So that in less than one year, if the need existed, Britain could swarm the seas with five hundred ships of war of all classes. Many of these are first-class battle ships of the line, and altogether they carry a complement of sixty or seventy thousand capable seamen. In addition to the heavy guns carried by the larger vessels, they are also provided with quick-firing guns of smaller pattern. The heaviest ordnance in the service are the four 81-ton guns, which are carried by the *Inflexible*. These figures and facts will give some idea of the enormous naval power of Britain. Well may she sing "Rule Britannia," and "Britons never will be slaves."

Having seen all we desired in Birkenhead dock-yards, and we were not so privileged for many a long day, having acquired information that it would be impossible to acquire elsewhere, we tendered our thanks to the authorities for their extreme courtesy, and returned to Liverpool to make preparations for leaving England.

So accordingly next day we said farewell to this merry land, every foot of which has been made sacred by events which have passed into history, and set sail, per Royal Mail Steamship *Vancouver*, for Ireland.

We have, in the former part of this work, given a description of the steamship *Labrador* that bore us from American shores to the Old Land, so now we take this opportunity of describing the majestic vessel which bore us homeward to our loved Canadian shores. She was 420 feet in length, 50 feet beam, drawing 26 feet of water, registering 5,600 tons, with four iron masts rigged with wire rope, and consuming 100 tons of coal daily. The vessel is specially constructed for the comfort and convenience of passengers, and is capable of steaming in the neighbourhood of fourteen knots per hour. This splendid steamer of the Dominion Line anchored the next day after our departure from Liverpool off Moville, in the north of Ireland.

CHAPTER XXII.

Ireland—Castle Green—Irish Politics, etc.

IMMEDIATELY after anchoring, boats from the shore came alongside, and a number of passengers, the writer included, clambered down the rope ladder into one of these and was rowed ashore. The moment we set foot upon the "old sod" we were besieged by about a score of jaunting-car drivers, who commenced in chorus to enlarge upon the merits of their respective horses and cars. One declared that he had the best outfit in all the land, and here we had the first evidence of native Irish wit, for we asked him if he always told the truth, and he at once replied, that he did except when a lie suited better. Before taking a car, however, we walked through the streets of the village to see what condition things were in, and may say regarding the matter, that if this was a true representation of such places in the Emerald Isle, the people are neat and thrifty, in the extreme; the streets were clean, the houses and shops were in the same condition, and the people looked neat and well-to-do.

We dropped into a cane store and purchased a few blackthorns for friends in Canada, and while there, several young Irishmen entered the place and showed us how they used the sticks on the fair-greens. They catch them in the centre and use both ends, and many a fine cranium has come to grief in the past under their smart blows. We, however, had no intention of putting any of them to such use, or presenting them to friends for such a purpose, and we are pleased to say that the practice is dying out amid Erin's bowers.

After completing our inspection of the village we went back to the shore and engaged a jaunting-car to carry us to an old ruin, some miles across the country. We asked the driver what he would charge for the round trip, and he, with consummate tact, said that he would leave it to our superior judgment, and, of course, placed upon our honour, we dare not do anything mean, but when we tendered him his fee were at the same time careful to evidence the superior judgment, which he had spoken of, by not making the sum unwisely large. We afterward discovered that some of the passengers had been charged extortionate sums for the trip, one young Englishman paying three times as much as the writer; and those who knew what we had paid considered the sum very respectable, and the driver seemed perfectly satisfied.

While in the city of London we were informed that some tourists engaging a conveyance have been outrageously fleeced, having previously arranged for the price per mile, and in many such cases being driven a round-about way. One in particular was brought to our notice, where the individual arranged with the hansom driver, for a shilling an hour, to drive to a certain point, and this functionary, not being particularly scrupulous, drove many miles around the city before starting for the point specified, and when they had returned, compelled the tourist to pay for every hour. Londoners informed us that Americans were more frequently taken in than Canadians; the former had the appearance and manner of sharp, clever people, but at bottom were not so smart as they appeared. On the contrary, the latter did not make any pretensions to cleverness, but it was found almost impossible to fool Canadians, so deep and shrewd were they. In all our travels in England and Ireland, we never noticed any disposition to take advantage of us. We do not agree with other travellers when they say they are fleeced at every turn, and that hotel land-lords, and 'bus drivers, and railway officials are all "sharks." If they found this to be the case it was not due to the grasping propensities of these persons so much as to their own excessive greenness.

Having now reached the ruins of the old castle, we proceeded to explore them, and wher-ever fragments of walls were standing, noticed that the masonry was of the first order; some of the arches were standing in their entirety as imposing as ever before in their history, others had crumbled into decay, and their fragments were now covered with moss. Altogether the ancient ruins presented a strange weird appearance, filling thoughtful minds with wonder as to their age, and so on inquiry concerning this, we were informed, that as far as can be determined they are more than a thousand years old, having been occupied by one of the ancient kings of Ireland, perhaps some time before the reign of Brian Boroimhe.

After viewing the remains of this historic pile, we had a most delightful drive along a quiet country road fringed with clumps of blackthorn bushes, and so elevated that it afforded us an excellent view of a vast tract of lovely landscape. During this drive we met a very fine-looking young lady, and asked the driver if all the Irish ladies were as good-looking as she, and he said they were better, adding, " indeed they are so handsome that young men come all the way from California to Ireland for wives." He then asked the author if he were a Holy Catholic, and we politely informed him that most fortunately we were a Protestant. He, however, did not manifest any discomposure at this, though a Roman Catholic himself. We asked him where the Methodist minister lived, and he pointed out the parsonage; we also propounded the question what manner of man he was, thus trying him, but with the shrewdness of a statesman he declared that the parson was a superior person.

We now said farewell to this witty, intelligent young man, who was exceedingly bright, and called on the divine, who, being presented with our card, received us with the greatest cordiality, as did also his excellent wife. After an interesting conversation, we took our leave, inviting the venerable pair to visit Canada at some future time, but the aged lady said, in beautiful language and evidently with longing desire, that she thought the first trip they would take would be to the heavenly country. This aged minister of Christ accompanied us through the town and informed us of the condition of Ireland, both religiously and politically—that in Donegal, upon whose soil we now trod, there were three Catholics to one Protestant. We asked him regarding Home Rule, if he thought it would be an advantage to Ireland, and in reply, with tears in his eyes, he said with terrible earnestness, "the Lord deliver us from Home Rule. It would mean Rome Rule and woe to Protestantism."

It is not the desire of the author, however, to make this work objectionable to Roman Catholics. Some of the most delightful conversations which we had, while eastward bound, were with a priest, who was principal of a Montreal college and a cultured and gentlemanly man; also, during our travels, we had received numerous kindnesses from others who belonged to the laity of the same Church, and consequently anything on our part that would reflect upon those who had been courteous and kind would be an unpardonable offence.

We noticed during our brief sojourn in this department of Her Majesty's possessions, members of the Royal Irish Constabulary, armed like regular troops of the line, and said by experts to be the finest-looking body of men on earth; also, there are in Ireland thirty thousand troops, some of them crack regiments of the British army.

CHAPTER XXIII.

THE BRITISH ARMY.

As we have given some figures relative to the strength of the navy, at an earlier period, it is now in place to give some facts regarding the strength of the military service, which has been suggested by the sight of some members of both the police and military forces in northern Ireland. The British army is supposed to be commanded by the Sovereign, assisted by the Secretary of State for War, in certain matters, and by the chief commanding officer in others. The elements of the force are, the rifles, volunteer artillery, reserve, cavalry, yeomanry, militia, native troops, engineers, regular artillery, ordnance corps, cavalry of the line, infantry of the line, household troops.

The latest official figures which we chance to possess are the following : Horse artillery, 5,609 ; cavalry, 17,219 ; artillery, 28,892 ; engineers,

5,626; infantry, 122,134; service corps, 2,990;
colonial corps, 2,485; army hospital corps, 1,745;
additional force, recruited for foreign service,
3,900; total, 190,600. In addition to the above,
we have the army reserve, of the first class,
22,000; second class, 24,000; militia, with
reserve, 137,556; yeomanry cavalry, 14,614,
volunteers, 244,263; grand total of all arms,
633,033. This force may appear small when
compared with the enormous standing armies of
the continent, but it must be remembered that
its discipline is perfection itself, and its members
are armed with the best weapons that human
skill and cunning can devise. The writer knows
whereof he speaks, having mingled with the
men of some of the best regiments and examined
their arms.

The artillery armament consists of a large
number of batteries, composed of powerful
12-pounder breech-loading rifled guns. Putting
the naval and military power of the Empire
together, we have nothing to fear from hostile
arms. Britannia loves peace, and trusts in the
God of peace, and thus, in the midst of threat-

ened danger, can look out over the darkening wastes and say, " Welcome the three corners of the world in arms, and we shall shock them ; " or, in Gerald Massey's soul-stirring words,

> "Rouse the old royal soul,
>   Europe's best hope
> Is Britain's sword-edge by victory set.
>   She shall dash freedom's foes
>   Down death's bloody slope,
> For there's life in the Old Land yet."

The End.